Integrated Water Environment Treatment

Integrated Water Environment Treatment

Mountainous Sponge City and Three Gorges Reservoir Region

Xiaoling Lei and Bo Lu

CRC Press
Taylor & Francis Group
Boca Raton London New York

CRC Press is an imprint of the
Taylor & Francis Group, an **informa** business

CRC Press/Balkema is an imprint of the Taylor & Francis Group, an informa business

© 2021 Taylor & Francis Group, London, UK

Typeset by codeMantra

Library of Congress Cataloging-in-Publication Data
Names: Xiaoling, Lei, 1967– author. | Bo, Lu, 1974– author.
Title: Integrated water environment treatment: mountainous sponge
city and Three Gorges Reservoir Region / by Lei Xiaoling and Lu Bo.
Description: Leiden, The Netherlands; Boca Raton: CRC Press/
Balkema, an imprint of Taylor & Francis, [2021] | Includes
bibliographical references and index.
Identifiers: LCCN 2020034991 (print) | LCCN 2020034992 (ebook) |
ISBN 9780367629434 (hardback) | ISBN 9781003131267 (ebook)
Subjects: LCSH: Watershed management—China—Yangtze River
Watershed. | Runoff—China—Yangtze River Watershed. | Urban
hydrology—China.
Classification: LCC TC502.Y3 X53 2021 (print) | LCC TC502.Y3
(ebook) | DDC 628.109512/12—dc23
LC record available at https://lccn.loc.gov/2020034991
LC ebook record available at https://lccn.loc.gov/2020034992

Published by: CRC Press/Balkema
               Schipholweg 107C, 2316 XC Leiden, The Netherlands
               e-mail: Pub.NL@taylorandfrancis.com
               www.crcpress.com – www.taylorandfrancis.com

ISBN: 978-0-367-62943-4 (hbk)
ISBN: 978-1-003-13126-7 (ebk)

DOI: 10.1201/b22623

https://doi.org/10.1201/b22623

Contents

Foreword

Water is the source of life, essential of production and foundation of ecology. After stepping into the new era of ecological civilization, China has upgraded sustainable development to green development through ecological civilization construction, because water resources are crucial to the sustained economic prosperity of China. As a hilly and mountainous country, China has a mountainous area of about 6.5 million km^2, with mountainous cities and towns accounting for about half of its total cities and towns, so the mountainous city construction plays an important part in the urbanization process of China. Chongqing has the characteristics of a typical mountainous city, such as the large terrain elevation difference, large road gradient, rapid convergence speed, early rainfall peak and serious underlying surface pollution. In consequence, the storm waterlogging risk management, peak runoff reduction and runoff pollution control in complex terrain have always been primary goals of water environment treatment. Integrated water environment treatment is a complex engineering issue. No single treatment measure can solve all the water environment problems, and the water environment treatment cannot be completed overnight. Therefore, a treatment model from multiple aspects and angles is needed to improve the efficiency of integrated water environment treatment.

The integrated water environment treatment shall be fully considered from the aspects of water safety, water quality improvement, water ecology, water landscape and water management. It not only helps to control the existing non-point source pollution and other problems but also can improve ecological functions of water environment systems. Meanwhile, the retention, storage and purification of stormwater runoff can accelerate the urban water environment treatment process, finally to achieve the water environment goal of long-term clear water, build a society with harmonious coexistence between human and nature and construct the urban ecological environment with clear water and beautiful scenery.

This monograph summarizes theories and practices of the integrated water environment treatment in Chongqing, a typical representative mountainous city in the Three Gorges Reservoir Region. Based on the mountainous terrain and unique water environment characteristics in the Three Gorges Reservoir Region, the monograph analyzes measures and difficulties in the basin water environment integrated treatment, and timely makes updates and improvements for drainage during the novel coronavirus outbreak. The ideas of keeping pace with the times and adapting to local conditions are worth learning and popularizing in integrated water environment treatment. This monograph, by taking the integrated water environment treatment of a mountainous city in the Three Gorges Reservoir Region as a practical case, introduces the integrated water environment treatment idea and relevant technical measures in detail and puts forward corresponding schemes by combining theories with practices, providing theoretical and technical support for integrated water environment treatment of mountainous cities.

Li'an Hou
Academician of the Chinese Academy of Engineering

Preface

Water resources are important resources to ensure the survival and life of urban residents. Urban rivers refer to the rivers or river segments originating from or flowing through urban areas with multiple ecological functions such as the supply of water resources, environmental protection, transportation, provision of green space, tourism, entertainment and cultural education. Urban rivers are important urban environmental factors affecting the formation and development of cities as important carriers of resources and environment. Meanwhile, urban rivers are also affected by human activities and urban development. Under the background of rapid urbanization and rapid economic development, a series of imbalanced relationships between water environment quality and economic development, water ecological balance and human activities, water environment capacity and pollutant discharge, natural purification and artificial treatment have led to water pollution problems of urban rivers.

Water environment is an essential condition for human survival and development. Basins are not only contradictory carriers of water resources and water environment problems, but also concentrated manifestations of the conflicts between water ecological environment protection and social and economic activities. Urban basins are one of the important ecological systems of cities, facing water problems such as water pollution, water shortage and flood disasters during the rapid urbanization. Besides, water resource pollution is always accompanied with serious ecological environment damage, restricting the social and economic development. Water environment problems caused by urbanization development are critical issues in the urban ecological construction.

Water environment problems are comprehensive issues involving land utilization, relations among basins of multiple water body types and multiple pollution types. Therefore, overall planning shall be made by applying the concept of regional water environment treatment according to the regional water environment characteristics and overall treatment goals, with full consideration to the systematicness, hierarchy and scale of the water ecosystem structures and functions to avoid making inconsiderate artificial design by simply splitting the water space from natural ecosystem. Systematic planning and design schemes and reasonable process technology are basic guarantees for successful water environment treatment.

Water environment problems are difficult to be treated and easily reoccur, seriously affecting the urban environmental and ecological safety, which have become one of the important factors restricting the sustainable economic and social development and ecological civilization construction in China. As the current urbanization rate of

China has exceeded 50%, the coming period will be a high risk period of urban water safety and also a critical opportunity period for water pollution treatment and water ecology restoration. The pollution control of urban rivers and lakes has become an essential condition and a key task to ensure the sustainable and healthy development of urban living environment, economy and society. At present, the water environment integrated treatment measures in many cities of China have achieved preliminary effects, but the water environment pollution problems are still very serious in many places. It can be predicated that various engineering measures and non-engineering measures for water environment pollution treatment will emerge one after another. Therefore, the vast environmental protection workers shall establish correct ideas of controlling water pollution, restoring water ecology and ensuring water security to realize healthy water circulation.

Located at the junction of the middle and upper reaches of the Yangtze River, the Three Gorges Reservoir Region, extending from Yichang City of Hubei Province in the east to Jiangjin District of Chongqing in the west, is a key development area for promoting the east-to-west economic development in the Yangtze River basin. As the largest mountainous city in the Three Gorges Reservoir Region, Chongqing is characterized by abundant water systems, fragmented land, intense human activities and relatively sensitive natural ecological environment. During its water pollution control process, the principles of water conservation first, decentralized treatment, in situ utilization and overall management shall be adhered to by reserving natural landforms such as the original ponds, streams, valleys and banks and intensifying the research, development, popularization and application of new process, new technology and new materials to effectively promote the integrated treatment of urban water environment.

Themed on the research and practice of typical mountainous city integrated water environment treatment in the Three Gorges Reservoir Region, this book systematically expounds the concepts of the Three Gorges Reservoir Region and mountainous city and clarifies the significance of the basin integrated treatment. Besides, taking the typical mountainous city Chongqing as an example for the integrated water environment treatment, it details the Chongqing basin water environmental problems and their causes, corresponding integrated water environment treatment strategies and typical measures taken, novel coronavirus prevention and control measures of drainage systems in mountainous cities, and results achieved through implementation of these measures. By sharing practical experience in integrated water environment treatment of mountainous cities, it provides some ideas for the integrated water environment treatment of China's cities, especially mountainous cities.

Due to the limitations of knowledge scope and academic level, this book may be subject to improvement, and your criticism and corrections are appreciated.

Abstract

Water is an essential condition for human survival and development. The rapid economic development accelerates the urbanization process of China, but the rapid growth of urban population and large-scale clustering of modern industries lead to the sharp increase in sewage discharge and water environment pollution. Meanwhile, the impacts of water pollution on environment also seriously threaten the human health. Water environment problems are difficult to be treated and easily reoccur, seriously affecting the urban environmental and ecological safety. In face of difficulties and problems in water environment treatment, reasonable integrated water environment treatment measures and strategies shall be explored to realize the harmonious coexistence between human and nature as soon as possible.

Integrated water environment treatment is a systematic project, which is hard to be realized simply through engineering measures, but also needs the cooperation of some non-engineering measures, policies and systems. The basin water environment is the main carrier of urban stormwater and also the receiving body of all kinds of pollution. Integrated water environment treatment helps control non-point source pollution and improve ecological functions of water environment, and the retention and purification of stormwater runoff through sponge city construction are important guarantees for achieving long-term clear water after the integrated treatment.

Themed on the research and practice of the mountainous city integrated water environment treatment in the Three Gorges Reservoir Region, this monograph points out the necessity of integrated water environment treatment from the aspects of current water environment situation and water pollution hazards, introduces the special geographical location and water environment characteristics of mountainous cities in the Three Gorges Reservoir Region and systematically introduces the necessity and significance of integrated water environment treatment in mountainous basins from multiple aspects and angles. The monograph details the water environment problems in Chongqing basin before integrated treatment and analyzes the causes of such water pollution by taking the typical mountainous city Chongqing as an example. Based on the analysis of water environment problems, it also introduces corresponding treatment measures and typical integrated treatment measures taken in the subsequent integrated water environment treatment and gives an in-depth introduction of concrete practical schemes for Chongqing sponge city planning, sponge community construction, sponge park construction and integrated water environment treatment of the Panxi River in Chongqing main urban area. Particularly, it also elaborates the prevention and control measures for urban drainage systems in mountainous regions during the novel coronavirus pneumonia outbreak in view of the current COVID-19 transmission. Finally, it describes the results archived by Chongqing in integrated water environment treatment and prospects for the future.

The monograph provides some ideas for the integrated water environment treatment of other cities around the world, especially mountainous cities, through sharing practical experience in integrated water environment treatment of mountainous cities and various prevention and control measures for drainage systems during the outbreak.

Acknowledgments

This book was written by Xiaoling Lei, a professor from Chongqing Academy of Science and Technology, and Bo Lu, a professor from Chongqing Municipal Engineering Technology Research Center of Sponge City Construction, by reference to the *Mountainous Sponge City Construction Series* edited by the two authors. It took 2 years to finish the book. Zhongsheng Pan, Jiawei Niu, Qiucheng Liu, Han Liang and Qin Xiao participated in the preparation and proofreading processes of the book. We would like to express our heartfelt gratitude to all the friends who made efforts for this book and also gratitude to financial support for the book from the following projects: Case study on implementation of "river chief system" in the Yangtze River Economic Belt, project No.: 2019-XZ-23; research on development strategies for Chongqing drinking water source pollution control and drinking water safety precautions, project No.: 2019-CQ-ZD-1; research on strategies for integrated water environment treatment in Chongqing basin, project No.: 2018-XZ-CQ-6. The above may not be exhaustive. Thanks again!

Authors

Xiaoling Lei, female, born in September 1967, professor, graduate adviser, member of Water Environment & Ecology Sub-society of Chinese Society for Urban Studies, editorial board member of China Water & Wastewater, listed in Chongqing Talents—Innovation Leading Talent and awarded the honor of "main force" in rural water environment of 2019, is the head of the Research Center for Low Carbon and Environmental Engineering of Chongqing Academy of Science and Technology, and concurrently holds the positions of the Deputy Director of Chongqing Municipal Engineering Technology Research Center of Sponge City Construction, Deputy Director of China-Canada Three Gorges Water Science Centre, Deputy Director of Chongqing Sponge City Construction Expert Committee, Vice President of Water Environment Institute of Chongqing Jiaotong University, etc. From 1986 to 1994, she studied at Tsinghua University majoring in environmental engineering, and from 2002 to 2005, she studied at University of British Columbia, Canada, majoring in environmental engineering. She has successively worked for Shenzhen Environmental Protection Bureau, Shenzhen Water Authority, Suez Sino-French Water Group (Macau), Chongqing Jiaotong University and Chongqing Academy of Science and Technology, and has long engaged in the industrial management, teaching research and technology research and development in such professional fields like municipal water supply and drainage, rural drinking water safety, water service operation and management, sponge city construction, water environment treatment, low carbon technology and carbon trading. She has presided over and completed more than 40 national, provincial and ministerial-level scientific research projects; led the establishment of respectively one national and municipal bi-lingual model course; won five provincial and ministerial-level science and technology awards; published over 80 papers in domestic and foreign academic journals; obtained 16 national patents and two software copyrights with her independently researched and developed core technologies; published four academic works and one audio-visual monograph; and participated in the compilation of one specialized textbook and more than ten local standards, such as Administrative Measures for Sponge City Construction in Chongqing, Code for Design of Outdoor Drainage Pipeline in Mountainous City and Standard Design Drawings for Low Impact Development Stormwater Facilities in Urban Roads and Open Spaces of Chongqing.

Bo Lu, male, born in February 1974, doctor, scientific and technological innovation leader granted by the MOHURD at the 70th Anniversary of the founding of the People's Republic of China, professor-level senior engineer, doctoral supervisor, now is the Director of Chongqing Municipal Engineering Technology Research Center of Sponge City Construction, Assistant President of Chongqing Municipal Research Institute of Design, Vice President (part-time) of Water Environment Institute of Chongqing Jiaotong University, Deputy Director of Chongqing Sponge City Construction Expert Committee, member of Senior Professional Title Evaluation Committee of Chongqing Urban Management Commission, one of the first-batch national registered public utility engineer (water supply and drainage), American registered project manager, technological leader of the Central Black and Odorous Water Inspectorate, Vice President of Utility Tunnel Construction and Underground Space Utilization Specialized Committee of China Municipal Engineering Association, member of Water System Engineering and Technology Branch of China Exploration and Design Association, member of Water System Intelligent Technology Research Institute of Water Industry Branch of China Civil Engineering Society and one of the first outstanding young designers of Chongqing.

Engaging in the exploration and design of environmental engineering projects for more than 20 years, he has presided over and completed more than 100 national, provincial and ministerial-level exploration and design projects; published more than 30 academic papers in domestic and foreign academic journals and at academic conferences; won more than 30 provincial and ministerial-level awards for scientific and technological progresses and excellent engineering exploration and designs; obtained the licensing for over 30 national invention and utility model patents and published three monographs on sponge city and one monograph on black and odorous water treatment.

He is also the chief editor or editing participant of many national and local standards, such as *Administrative Measures for Sponge City Construction in Chongqing, Guideline for Sponge City Planning and Design of Chongqing, Code for Design of Outdoor Drainage Pipeline in Mountainous City, Standard for Design of Low Impact Development Stormwater System* and *Deep Provisions for Preparation of Engineering Design Documents for Sponge City Construction in Chongqing.*

Chapter I

Introduction

1.1 Overview of integrated water environment treatment

With the rapid development of industrialization and urbanization, the water pollution control and water environment treatment not only gradually become major issues of global concern but also are major constraints to the healthy, rapid and sustainable economic and social development of China. The imperfect environmental infrastructure, incompletely effective treatment of urban domestic pollutants, improvement-needed industrial pollution treatment and gaps in agricultural pollution control lead to the poor water quality, water eutrophication, and even black and odorous water or water quality function decline of some water bodies. The current water environment is not optimistic. We are still struggling to cope with the water resource scarcity problem and ecological destruction phenomenon that affect the human life and harm the human health. The water environment treatment mainly starts from the following five aspects:

1. Promote agricultural modernization and reduce agricultural pollution. Accelerate the agricultural development mode and actively develop agricultural circular economy to develop the green agriculture.
2. Improve the urban infrastructure and reduce the domestic pollution. With the increasing consumption of urban domestic water and discharge of domestic sewage due to the increased urban population density, the urban drainage pipes, pipe networks and domestic sewage plants shall be constructed and optimized synchronously to continuously improve the water environment.
3. Realize wastewater recycling and reduce industrial pollution. The industrial wastewater discharge keeps increasing with the development of economy. The government shall tighten its grip on the industrial wastewater to strictly prohibit the excessive sewage discharge or the discharge of sewage not meeting the treatment standard, so as to realize the transformation of wastewater resource and increase the utilization rate of water resource.
4. Improve the environmental monitoring technology level and enhance the pollution accident response capacity. Improve the technical equipment at basic-level environmental monitoring stations to enhance the pollution accident response capacity as well as the prediction and early warning abilities, monitor the dynamic changes of pollution sources and increase the monitoring frequency to reduce the sampling error, so as to timely reflect the pollution discharge situation and enhance the environmental management capability.

5. Promote the construction of sponge city. The purpose of sponge city construction is to deal with the relation between urban construction and protection of water resources and ecological environment, which reflects an upgraded understanding of the urban concept and the urban water demand. The large-scale and rapid urbanization process changes the underlying conditions and even the landforms and source water system, thereby changing the natural hydrological characteristics, such as the original evaporation, infiltration, slope runoff generation and convergence. Subsequently, the sharply reduced urban detention storage capacity will lead to a series of problems, such as the loss of stormwater resources, increase of runoff pollution and frequent waterlogging. The construction of sponge city is adopted with corresponding technical measures for the "infiltration, retention, storage, purification, utilization and discharge", focusing on improving the urban water ecological environment and building a benign water circulation system to make the city greener, more ecological and livable. The stormwater runoff control is an important content in the construction of sponge city, usually realized by building storage ponds by means of natural water, multi-functional water storage and artificial regulation. Through stormwater retention, storage and purification, the sponge city ecological engineering can increase the water flow and environmental capacity, help the non-point source pollution control and prevent black and odorous water occurrence, which is an important guarantee for the long-term clear water.

1.2 Necessity of integrated water environment treatment

1.2.1 Current situation of water environment

After the water impoundment in the Three Gorges Reservoir Region, the pollution range further expands due to the slower flow velocity in the Reservoir Region and tributaries. Due to combined effects of the flow velocity, temperature and pollutant discharge, the eutrophication of some tributaries in the Three Gorges Reservoir Region is getting severer.

Domestic sewage is one of the major water pollution sources of the Three Gorges Reservoir Region. Although urban domestic sewage treatment facilities in the Reservoir Region have been gradually built, part of the sewage is still discharged into water bodies due to the imperfect secondary and tertiary sewage collection networks and incomplete separation of rain from sewage.

One of the major causes of water environment pollution in the Three Gorges Reservoir Region is the non-standard vessel sewage treatment and discharge to the Yangtze River.

Inadequate water self-purification capacity is one of the main causes of water environment pollution in the Three Gorges Reservoir Region. Considering the weak water environment self-purification capacity and very high total discharge of pollutants in the Three Gorges Reservoir Region, it is difficult for the accumulated pollutants to be degraded merely by river runoff.

In general, since the construction of the Three Gorges Project, China has always attached great importance to the environmental and ecological protection and construction in the Reservoir Region. Authorities concerned and all departments of the

Reservoir Region have increased input and strengthened the management and control, making remarkable results in the water pollution control in the Reservoir Region. The water environment in the Three Gorges Reservoir Region is being improved gradually.

1.2.2 Hazards of water environment pollution

Water environment pollution reduces the water quality of water resources in China and affects the sustainable economic development, and also directly relates to the drinking water safety, endangering the public health and life safety.

Polluted water can directly or indirectly affect the human health. Pathogenic bacteria in dirty water can invade the human body and cause diseases, such as schistosomiasis and leptospirosis, while physical and chemical pollution may cause irreversible impacts on human genetic material.

As for agriculture and fishery, the irrigation of farmland directly with sewage containing toxic and harmful substances will pollute the farmland soil, destroy the original good soil structure, reduce the crop quality and yield and eventually lead to crop failure. Especially in arid and semi-arid areas, the sewage irrigation may increase the crop yield in the short term, but toxic substances will accumulate in vegetables and grains, threatening the human life. Besides, water pollution may lead to a sharp decline in the quantities or even extinction of fish and aquatic life in the water. Meanwhile, water pollution will also reduce the output of freshwater fishery and mariculture.

As for industrial production, water pollution may increase the industrial water consumption, resulting in increased costs. As for edible industrial water, the process inspections become more rigorous, causing a series of consequences, such as reduction of industrial production capacity and decline in output, thus affecting the normal growth of enterprises.

As for the water eutrophication, the discharge of nitrogen, phosphorus, potassium and elements leads to a sharp increase in organic content in water, providing favorable conditions for the survival of algae. The sharply increased algae cover the original clean water surface and prevent oxygen in the air from dissolving in water. Consequently, the oxygen content of water drops sharply, resulting in the death of aquatic life. Due to relatively high chemical oxygen demand (COD) of main pollutants and content of nitrogen, phosphorus and other nutrients in sewage, the eutrophication may also occur in the surface water of lakes and seas in varying degrees.

1.2.3 Development prospect of integrated treatment

With the accelerated urbanization and increasing urban population of China, the sewage discharge and water environment pollution have also increased sharply. The integrated treatment of sewage in the water environment is urgently needed, which shall keep pace with the times to improve the environment and reduce the waste of water resources. Besides, the scientific solution to integrated sewage treatment is also an important part for realizing social sustainable development, lucid waters and lush mountains.

The traditional water environment treatment usually adopts the unitary mode of "end treatment" of the river treatment. The "end treatment" is a short-term solution that "only addresses the symptoms not the causes". The scale of investment in an end treatment project is small, generally less than nine figures, not enough for the overall purpose. The environmental goals have been overfulfilled in the early stage, but the overall water environment in each basin and urban river has not been obviously improved. After issuance of the "Ten Measures for Water Pollution Control" and other planning, the idea has been shifted from the traditional way of "end treatment" to the whole-process water pollution control mode of "pollutant discharge reduction at the source, pollutant blocking in the process and end treatment", and the "systematic mode" featured with basin environmental units consisting of pipe networks, sewage treatment plants and rivers has also been established gradually for the environmental governance projects.

1.3 Typical mountainous city in Three Gorges Reservoir Region

1.3.1 Three Gorges Reservoir Region

1.3.1.1 Concept of Three Gorges Reservoir Region

Due to the construction of the Three Gorges Hydropower Station in Yichang, Hubei Province at the end of the Yangtze River Three Gorges, some cities and towns would be submerged due to the rising water level in areas upstream of the reservoir, so the mass migration was needed. Under such background, the Three Gorges Reservoir Region was formed, including 20 districts and counties needing mass migration due to the rising water level caused by the Three Gorges Project construction. With geographical coordinates of 106°20′–110°30′ E and 29°–31°50′ N, the Three Gorges Reservoir Region covers Yichang County, Xingshan County and Zigui County under the jurisdiction of Hubei Province; Badong County under the jurisdiction of Enshi Prefecture; Wushan County, Wuxi County, Fengjie County, Yunyang County, Kaixian County, Wanzhou District, Zhongxian County, Fuling District, Fengdu County, Wulong County, Shizhu County, Changshou County, Yubei District, Baxian County, Jiangjin City and Chongqing main urban area (including Yuzhong District, Nan'an District, Shapingba District, Dadukou District, Jiulongpo District and Jiangbei District) under the jurisdiction of Chongqing, including the Yichang–Chongqing watershed area along the Yangtze River, with the submerged land area of about 632 km².

1.3.1.2 Geographic and geomorphic conditions of Three Gorges Reservoir Region

Located at the junction of the middle and lower plains of the Yangtze River, the Three Gorges Reservoir Region covers the low mountain and valley areas of Sichuan and Hubei, East Sichuan paralleled ridge and valley area as well as low mountain and hilly area, with Daba Mountains on the north and Yunnan–Guizhou Plateau on the south. The terrain of China is divided into three levels and the second one mainly includes plateaus and basins. The Three Gorges Reservoir Region is located in the slope area on the second level and forms a great terrain elevation difference with the adjacent Hanjiang Plain on the third level, resulting in the river downcutting, which makes the Reservoir Region form the geomorphic pattern of fractured mountains and rugged terrain full of ravines. Most areas of the Reservoir Region are characterized by high mountains and deep valleys, with the hills and mountains accounting for 95.7% of the total area, among which the mountains account for 74% of the total area, while the flat area accounts only for 4.3% of the total Reservoir Region area.

The Three Gorges Reservoir Region mainly experienced the Jinning movement prior to the Sinian Period, the Yanshan movement at the end of the Jurassic Period and the Himalayan movement at the end of the Tertiary Period. The three tectonic movements formed strong folds and faults in the strata of the Reservoir Region, then the Reservoir Region entered the neotectonic period, characterized by the overall topography uplift. Due to the sea–land changes and water downcutting caused by strong geological movements, unique ravine landforms formed in the Reservoir Region and a relatively independent geographic unit was also formed. The overall terrain gradually slopes downward from the basin edge to the center. The elevation in Fengjie is about 1,000 m and the elevations of towering mountain ridges are usually about 700–800 m, gradually down to gentle hills in Changshou with elevations about 300–500 m. A flat broad valley is formed when the Yangtze River winds in a synclinal valley, and a small gorge is always formed in a cross-cutting and anticlinal section. The high mountains and steep bank slopes on both banks of the Yangtze River as well as the narrow valleys and rapid flow form the famous Three Gorges of Yangtze River.

1.3.1.3 Characteristics of water environment in Three Gorges Reservoir Region

The Three Gorges Reservoir Region in the middle and upper reaches of the Yangtze River is located in a transitional zone between the south temperate zone and the subtropical zone, and has a subtropical monsoon climate, with the annual average precipitation of 1,100–1,200 mm and rainfall evenly distributed. Due to the expanded water area and increased water evaporation after impoundment of the Three Gorges Reservoir, the day-and-night temperature difference in the Reservoir Region is reduced, changing the climatic environment in the Reservoir Region. The natural river channel in the Reservoir Region has become a reservoir, causing significant changes in hydrological characteristics of the main stream and many tributaries of the Yangtze River. The water environment in the Reservoir Region has gradually evolved from the river ecosystem of lotic environment to the lake ecosystem of lentic environment, and the migration and transformation ways and retention time of nitrogen, phosphorus and other nutrients have been changed significantly. As a result, the water eutrophication of tributaries is gradually aggravated, and there are frequent outbreaks of algal blooms in backwater areas of some tributaries. Meanwhile, with the rapid economic development in areas upstream of the Reservoir Region, the pollution load entering the Reservoir Region is constantly increasing and the deterioration trend of water environment in the Three Gorges Reservoir Region is still not relieved.

1. Point source pollution problems.

    The point source pollution mainly includes the pollution of industrial wastewater and urban domestic sewage, which are usually discharged into water bodies through fixed sewage outfalls. As for Chongqing main urban area, the point source pollution mainly has the following characteristics:

    ①   There are relatively less industrial enterprises in Chongqing main urban area, and the management on sewage discharge of industrial enterprises is relatively strict that the industrial wastewater can only be discharged into the urban sewage pipe network after being treated and meeting relevant discharge standards. Therefore, the point source pollution in Chongqing main urban area is mainly domestic pollution.

②　The sewage network coverage rate has not reached 100% yet, so the sewage in some areas is still discharged in an unorganized way or discharged into storm sewer.

③　Some old sewage pipes have the problem of serious leakage.

④　The treatment capacity of existing sewage treatment plants is limited.

Therefore, part of the domestic sewage is discharged into water bodies due to the inadequate sewage collection and treatment facilities, resulting in certain pollution to the surrounding water environment.

2. Rural non-point source pollution.

Chongqing Three Gorges Reservoir Region has a large rural area and a large agricultural population. The water quality in the Reservoir Region has been affected by the wide range of agricultural pollution, relatively backward pollution control technology, breeding pollution and crops fertilization for a long time.

3. Water body pollution problems.

The water quality of the main stream in the Reservoir Region has been improved gradually, but the pollution of some tributaries is still not optimistic. The result of assessment on the Yangtze River and Jialing River is as follows: the Yangtze River and Jialing River belong to Class III and Class II water bodies, respectively, with good water quality.

1.3.1.4 Ecological characteristics of Three Gorges Reservoir Region

Ecological environment is a limiting factor and material basis for the human survival and development, which is closely related to the people's life and production. The Three Gorges Reservoir Region is the most special ecological function zone of China and even the world, and its functions such as conservation of water and soil, water quality protection and maintenance of biodiversity are of important strategic significance for the long-term safe operation of the Three Gorges Project and the flood control and ecological safety in the middle and lower reaches of the Yangtze River. Since the impoundment of the Three Gorges Reservoir, a series of eco-environment improvement measures have been implemented gradually, but the eco-environment of Chongqing Three Gorges Reservoir Region has not been fundamentally improved and the current primary eco-environmental problems are mainly reflected in the following aspects:

1. The water and soil loss remains unsolved fundamentally, and the situation of land rocky desertification is grim. There are various land types in the Three Gorges Reservoir Region, characterized by large area of hills and mountains, small area of flat lands, complex land structures and significant vertical differences. Due to the distinct geographical conditions, there are frequent disasters such as earthquake, collapse, landslide and debris flow, and also serious water and soil loss. In addition, the massive loss of water and soil is also caused by the increase in population and advancement of urbanization.

2. The biodiversity protection and bio-safety issues have become increasingly needed for awareness. After the impoundment of the Three Gorges Reservoir in 2003, a large number of endangered and vulnerable species in the Reservoir Region are still vulnerable to the intervention of human activities. At present, the issues confronted with include the ecosystem degradation, species endangerment, loss of genetic resources, irrational exploitation of biological resources and protection work misunderstanding. Currently, the bio-safety issues in Chongqing Three Gorges Reservoir Region are increasingly apparent. If the supervision and management are not strengthened, the invasion of alien species may become an important factor that affects the environmental safety of the Reservoir Region and brings inestimable harm to especially the aquatic life in the Three Gorges Reservoir Region.

1.3.2 Mountainous city

1.3.2.1 Concept of mountainous city

The mountainous region can be interpreted in a broad sense or a narrow sense. In the narrow sense, it includes low mountains, middle mountains, high mountains and extreme high mountains, while in the broad sense, it includes mountains, hills and plateaus, and is the region with complex structure, complete ecological functions, diverse ecological processes and great influence in the earth surface system. According to this definition, the total area of the mountains, hills and rugged plateaus in China accounts for about 69% of China's total land area.

In engineering, the mountainous city is defined on the basis of the concept of geomorphology in the geography, characterized by the geomorphology of urban land and identified based on the terrain's impact on urban environment, urban engineering technology economy and urban layout. The nature of mountainous cities can be roughly summarized into three aspects: (1) geographical location. Most mountainous cities are located in large mountainous areas or at the junction of mountains and plains. (2) Social culture. The economy, ecology and social culture of mountainous cities form integral organic parts of the mountainous environment in the process of development. (3) Spatial characteristics. The topographic conditions that affect the urban construction and development have the complicated vertical geomorphologic characteristics of mountains unable to be overcome for a long time, thus forming the unique human residential space environment of terraced settlements and vertical differentiation.

China is a hilly and mountainous country with a mountainous area of about 6.5 million km^2 and mountainous cities and towns accounting for about half of China's total cities and towns. Mountainous cities and towns have a quite long history of formation and development, and the corresponding mountainous environment also has a stronger impact on the lives of contemporary people than that in the past. Mountainous cities and towns are not only the important places and components of the production and life of residents in mountainous areas but also important bases for the development of economy, society and culture.

Take Chongqing, the largest mountainous city in China, as an example. Chongqing is located in the paralleled ridge and valley area in east Sichuan Basin, with anticlinal low mountains Zhenwu Mountain and Zhongliang Mountain, respectively, on the east and west sides of its urban area. The urban areas are located in a low complex synclinal valley formed by these two mountains. The river surface in the urban area is 160–170 m above sea level, which is 300–400 m lower than the low mountains on both sides. The urban areas mainly are tablelands, hills and terraces, with large elevation difference and significant stratified structure. All buildings in the city are built skillfully on the mountains and scattered in a well-arranged manner, forming the mountainous city characteristic of "the mountain is a city, and vice versa". Chongqing is rich in river systems, with major rivers such as Yangtze River, Jialing River, Wujiang River, Fujiang River, Qijiang River and Daning River flowing through. Covering an area of 82,400 km², Chongqing has the jurisdiction over 38 districts and counties (autonomous counties), among which the built-up main urban area is 650 km².

1.3.2.2 Geographic and geomorphic conditions of mountainous cities

The rugged terrain, varied landforms and complicated natural ecological environment of mountainous cities bring great difficulties to the construction of mountainous cities, especially the planning and construction of urban infrastructure and traffic facilities, which are the main restraints on the construction and planning of mountainous cities. The complicated topographic, geomorphic and geological conditions of mountainous cities are mainly characterized by the following four aspects.

1. Land shortage and small road section. Due to the rugged terrain and land shortage, mountainous cities always have dense buildings and complicated underground pipelines for water, electricity, gas, communication, etc., resulting in less underground space available.
2. Complicated situation of urban underground utilization. Due to the complicated geographic and geomorphic conditions, many rails, tunnels and underpasses are usually built to alleviate traffic jams; and in order to make full use of the space, some underground stores, parking lots and air-raid shelters are also built leading to the complicated underground space utilization situation and deficiency in systematic planning.
3. Frequently and greatly undulating urban road surface. Compared with plain cities, mountainous cities have complicated geological, topographic and geomorphic conditions, greatly undulating terrain and steep slopes in some area, which are unfavorable for the layout of gravity drainage pipeline.
4. High hardening rate of urban underlying surface leading to waterlogging once it rains. Take Chongqing as an example. With the continuous expansion of the metropolitan area, the area of impermeable roads and squares in the built-up area keeps increasing. Although some stormwater can be discharged through municipal drainage pipes, it is difficult to solve the drainage in low-lying areas, leading to frequent waterlogging disasters.

1.3.2.3 Hydrological and climatic conditions of mountainous cities

Due to the unique topographic conditions and climate types, mountainous cities have both hydrologic dynamics and climatic variability, specifically reflected in the following aspects:

1. Hydrologic dynamics. Chongqing is a city with many mountains and rivers, where the water for production and life and urban stormwater in mountainous areas flow into water bodies of the mountainous city through various channels and the water bodies become the receiving terminals of runoff.
2. High ecological sensitivity of water bodies of mountainous cities. Many small natural depressions and ponds are formed in mountainous cities due to the terrain, which easily become seasonal water bodies. They are very vulnerable to pollution because of their lack of connectivity with each other and high ecological sensitivity.
3. Difficulty in groundwater replenishment of mountainous cities. There are often rocks of poor infiltration capacity beneath the topsoil of mountainous cities. The water in the soil is collected in the lower part of the ground and cannot continue to infiltrate and replenish the groundwater, so the mountainous cities are characterized by poor groundwater and high water level.
4. Climatic variability. Special local climate is formed in mountainous cities due to the unique height and topography. In addition to the natural climate formed under the effects of longitude, latitude, sea and land, many microclimates are also formed and the variable climate will lead to the uneven distribution of rainfall. Meanwhile, the vertical climate change will lead to irregular rainfall and mountainous cities have large vertical elevation differences, so mountainous cities have great vertical climate differences and the phenomenon of "four seasons in one mountain and different weather within 10 km" is not rare.

1.3.2.4 Sewer system of mountainous city

The urban drainage system refers to the system of engineering facilities for treating urban sewage and stormwater, which is also an important link to realize "source control and discharge reduction" of urban pollution, usually consisting of drainage pipes and sewage treatment plants. As for sewage, the current urban drainage system uses pipelines to collect target sewage and then centralizedly treats and discharges it in areas downstream of urban rivers or estuarine areas; as for stormwater, the urban drainage system adopts the quick-drainage method to drain stormwater out of the urban area as soon as possible to reduce its impact on the city. Such way of centralized treatment and fast discharge not only reduces the impact on the local urban life but also gradually causes a lot of negative issues, for example, water resource shortage, water environment pollution, water ecological destruction and water safety. It indicates that the current drainage system is no longer suitable for the current sustainable development goals of urban and social resources. Most of China's cities were adopted with the direct-discharge combined drainage system in the early construction and most of them have been reconstructed into intercepting combined systems or separate systems with China's increasing emphasis on the environmental pollution prevention and control; newly built areas are mostly adopted with separate rain and sewage systems, and the drainage systems in many cities of China have formed the hybrid drainage systems with coexisting combined and separate systems.

Natural terrain is an important factor that affects the drainage of mountainous cities. Due to the complicated natural base and relatively large terrain elevation difference, mountainous cities are usually divided into different zones in the drainage planning based on the elevations, and then proper drainage systems are developed according to the specific conditions of each zone. In mountainous cities with steep slopes and riverbeds of large longitudinal gradients where the water flow in gullies is significantly affected by seasonal changes, the principle of draining to nearby areas at the same elevations is adopted for urban storm sewer, and the drainage and flood discharge channels shall be arranged in a smooth and straight manner as much as possible to discharge water in rivers nearby, and in the long term, split-flow wells will be added to collect the initial stormwater runoff to sewage treatment plants. In order to prevent the impact of floods, cutoff ditches are designed to be set at the back edge of each drainage zone to channel the urban trailing stormwater into gullies nearby to prevent water from flowing to urban area, so as to alleviate the burden of urban drainage system; longitudinal open channels or slab culverts for flood discharge are set at each gully in urban areas to discharge floods directly into rivers. Horizontal storm sewers are set in urban areas along contour lines to discharge stormwater into longitudinal open channels or slab culverts for flood discharge. The drainage networks of mountainous cities usually have the following characteristics:

1. Complicated terrain. Mountainous urban areas are generally characterized by undulating terrain, large road gradients and serious segmentation of land. Usually forming tablelands at different elevations are formed, with relatively large ground elevation differences among them.
2. In the planning and design processes of sewage network, the large terrain elevation difference makes it more difficult to set sewage lift pump stations.
3. The complicated geological conditions featured with interlaced valleys and hills, large fill thickness and numerous scarps and steps raise the requirements for the safety and stability of pipes.

In recent years, most cities and towns have faced the plight of urban waterlogging after heavy rain in flood season. The problems like urban waterlogging and water pollution have caused damage to urban functions, social order, resources and environment to different degrees, which have become major issues in the economic and social development. Under the dual influence of geographic and geomorphic conditions and urbanization development, the rainfall confluence speed is fast in mountainous cities and a massive amount of stormwater runoff flows into municipal drainage network in a short time, which may lead to the instantaneous load of drainage network exceeding the designed drainage capacity. Especially in low-lying areas of mountainous cities, the stormwater runoff is apt to accumulate. If the accumulated stormwater is not timely discharged, it is very likely to cause stormwater backflow and even form urban flood, causing waterlogging disasters. Therefore, the drainage safety of mountainous cities is facing greater challenges.

1.3.2.5 Stormwater management system of mountainous cities

There have been more than 100 urban stormwater and flood management modes at home and abroad since the 1970s. Traditional urban development modes are mainly reflected in the increase in impermeable area and the efficient urban drainage system of "drainage network + sewage treatment + tail water discharge". On the one hand, the traditional fast discharge mode has achieved good effects on the pollution control, discharge reduction and urban water environment protection; on the other hand, it also has a great impact on the urban ecological environment.

The stormwater and flood management in China originated in the 1980s. Due to the rapid development of urbanization in China after the reform and opening up, the natural hydrological processes of cities were seriously damaged, and the flood disasters caused increasing erosion and threat to cities. In such context, many scholars began to explore stormwater management method suitable for China. After decades of research and development, a set of stormwater management network suitable for China's national conditions has been formed gradually, and a series of laws and regulations have been promulgated to ensure the smooth operation of stormwater management. The Ministry of Housing and Urban-Rural Development (MOHURD) of the People's Republic of China has issued drainage standards for various infrastructure to control the stormwater and flood management at a professional level. In October 2014, the *Technical Guideline for Sponge City Construction: Low Impact Develop Development Stormwater System Construction (Trial)* prepared by the MOHURD was officially released and implemented, providing theoretical guidance for the construction of sponge city. The *Technical Guideline for Sponge City Construction* specifies the connotation and construction method of sponge city, which provides practicable theoretical guidance for the stormwater management in China.

At present, China still has no comprehensive study on and application of stormwater management for mountainous cities. Many scholars have studied the stormwater and flood management in mountainous cities from different angles and levels. Professor Huang Guangyu mainly studied the causes, prevention and control of torrential floods in mountain cities in his book *Theory of Mountainurbanology*; in the article *Reflections on Waterlogging Control and Stormwater Utilization in Mountainous Cities*, Zhang Zhi analyzed the waterlogging and rainfall characteristics of mountainous cities and proposed the idea on the construction of integrated waterlogging prevention and stormwater utilization system in mountainous cities; in the article *Research and Application of Drainage System Model for Mountainous Cities*, Zhi Yue developed a drainage system model for mountainous cities based on the geographic information system (GIS) platform and combing with foreign storm water management model (SWMM) software. However, these studies usually only focused on a certain aspect of the stormwater management in mountainous cities, mostly from the perspective of engineering and lacking of systematic research on the stormwater management system for mountainous cities from the macro- to micro-perspective in combination with engineering and ecological facilities.

1.3.3 Chongqing—City of Mountains

The Three Gorges Reservoir Region is located at the junction of the middle and upper reaches of the Yangtze River, where mountainous cities account for more than half of the total cities. Chongqing is the largest mountainous city in the Three Gorges Reservoir Region and also the key economic development belt in the Yangtze River basin. Chongqing is historically characterized by large ground gradient, rapid convergence speed, short catchment time, abundant water systems, serious pollution of underlying surface and relatively sensitive natural ecological environment. However, since the impoundment of the Three Gorges Reservoir, the self-purification capacity of water body has reduced and the estuary sections of primary tributaries have been seriously polluted under the effect of the backwater. Under such unoptimistic situation of water environment, the management of storm waterlogging in complex terrain and the control of water environment pollution are the keys to the urban ecological construction of Chongqing.

1.3.3.1 Region characteristics

Located in the southwest of inland China and the upper reaches of the Yangtze River, Chongqing is a multi-center cluster-type city. The areas under the jurisdiction of Chongqing are mainly distributed along the Yangtze River, covering many hills and low mountains with an average elevation of 400 m, bordering Hunan, Hubei, Guizhou, Sichuan and Shaanxi Provinces. Chongqing is built leaning the mountains, known as "the City of Mountains"; due to its light rain and dense fog in winter and spring, Chongqing is also known as "the City of Fog"; the Jialing River in Chongqing was called Yushui in ancient times, so Chongqing is called "Yu" for short. Chongqing was approved as the fourth municipality directly under the central government of China at the Fifth Session of the Eighth National People's Congress on March 14, 1997, becoming the only municipality directly under the central government in the central and west China.

1.3.3.2 Topographic and geomorphic characteristics

Located in the east Sichuan Basin and the zone of transition to mountainous areas around the basin, Chongqing mainly consists of mountainous areas and hills and is the largest mountainous city in the Three Gorges Reservoir Region with its mountainous areas accounting for 76%. Surrounded by Dalou Mountain, Wushan Mountain, Wuling Mountain and Daba Mountain and with large slope area, Chongqing is known as "the City of Mountains". Due to the complicated terrain, diverse geologic structures are shaped by the interlaced stratigraphic structure systems in Chongqing, forming the undulating terrain and distinct stratification.

1. Greatly undulating terrain and distinct stratified landforms. The lowest elevation of Chongqing is 73.1 m, located in Yuxikou, Beishi Village, Wushan County, and the highest elevation is 2,797 m, located in Wushan, Wuxi. The east, southeast and south parts of Chongqing are mostly at elevations above 1,500 m, while the west part mainly consists of hills at elevations between 300 and 400 m.
2. Diverse landforms, mainly mountains and hills. The landforms of Chongqing are mainly divided into eight types of four categories: mountains (middle and low mountains), hills (high, middle, low and gentle hills), tablelands and flatlands, with mountains accounting for 75.8% and hills accounting for 18.2%.
3. Distinct differences of regions divided by landforms. The landforms are hills to the west of Huaying Mountain–Bayue Mountain; paralleled ridge and valley areas between Huaying Mountain and Fangdou Mountain; middle mountains of Daba Mountain in north Chongqing; and mountainous areas of Wushan Mountain and Dalou Mountain in the east, southeast and south Chongqing.
4. Widely distributed karst landforms. The typical karst landforms are widely distributed in east and southeast Chongqing, including stone forests, peak forests, depressions, gentle hills, ponors, karst caves, underground rivers and canyons, etc.

1.3.3.3 Climatic characteristics

The main climatic characteristics of Chongqing can be summarized as: warm winter and early spring, hot summer and cool autumn, four distinct seasons, long frost-free period; humid air, abundant rainfall; weak solar radiation and short sunshine duration; frequent cloud and mist, scarce frost and snow; light, heat and rainfall in the same season, distinct three-dimensional climate and abundant climate resources.

Chongqing has a mild climate, enjoying the subtropical humid monsoon climate with an annual average temperature of 16°C–18°C. The lowest temperature is in January, with the monthly average temperature at 7°C and the minimum temperature at −3.8°C. The highest temperature is in July and August, mostly between 27°C and 38°C, and the maximum temperature may reach at 43.8°C. Therefore, Chongqing, together with Wuhan and Nanjing, is known as the three "furnaces" in the Yangtze River basin. From the end of autumn to the beginning of spring, Chongqing is characterized by its foggy weather, with the annual average foggy days of 68 days, so it is also known as "the City of Fog".

1.3.3.4 Hydrological characteristics

The rivers flowing through Chongqing mainly include the Yangtze River, Jialing River, Wujiang River, Fujiang River, Qijiang River and Daning River. The main stream of the Yangtze River flows across Chongqing from west to east, about 665 km long, and runs across three anticlines of Wushan Mountain, forming the famous Qutang Gorge, Wuxia Gorge and Xiling Gorge (in Hubei Province), namely the well-known Three Gorges of Yangtze River. Jialing River merges into the Yangtze River in Yuzhong District and Wujiang River mergers into the Yangtze River in Fuling District.

The survey shows that the overall water resources available in Chongqing can basically meet the needs of water supply, but the spatial distribution is uneven. With abundant annual average precipitation and annual average relative humidity of 70%–80%, Chongqing is a high-humidity area in China, with 1,000–1,350 mm precipitation in most areas. The precipitation mainly occurs from May to September, accounting for about 70% of the annual total precipitation. The night rain in mountains is an important characteristic of Chongqing climate, and the stormwater runoff in Chongqing is characterized by short convergence time, rapid speed, strong erosion force, etc.

1.3.3.5 Drainage characteristics

Chongqing has the typical geographic and geomorphic conditions of mountainous cities in the Three Gorges Reservoir Region, namely undulating terrain and large terrain elevation difference, which makes the water flow in Chongqing have large potential energy difference and strong liquidity. Due to the complicated and varied geological conditions, the use conditions of the drainage network are complicated. Due to the complex terrain of cities and towns and greatly varied urban road gradients, the setting of drainage pipes is restricted by the terrain, overpasses, underpasses, other structures, traffic safety facilities and other factors, resulting in frequent urban road waterlogging that seriously affects the traffic. The discussion and research on the planning, design, construction, management and maintenance of the drainage system in Chongqing are needed to make corresponding optimization design.

In recent years, Chongqing has experienced the process of rapid urbanization. The urban land utilization has changed the urban surface water environment by changing the flow of material and energy, and its development and evolution have had a profound impact on the water environment. The increasing urban rainfall has caused problems like the loss of water and soil erosion. Although some stormwater can be discharged through the existing drainage network, the waterlogging often occurs in low-lying areas due to untimely drainage, easily leading to traffic jams, travel difficulties of residents, etc. Considering the characteristics of large ground gradient and strong water erosion force in mountainous cities, the runoff generated is seriously polluted, and the storm runoff in mountainous area leads to the more remarkable rapid occurrence and disappearance of changes in the basin hydrological processes, causing relatively strong impact on the urban water ecosystem. The non-point source pollution of high abruptness and strong impact caused by stormwater runoff has become one of the important reasons for the water environment deterioration. From the above, it is extremely urgent to carry out the planning and construction of mountainous sponge city in Chongqing.

1.4 Integrated basin treatment

1.4.1 Overview of integrated basin treatment

With the continuously rapid economic and social development and sharp increase in population of Chongqing, the urban construction land has expanded continuously and the river water supplement has decreased, resulting in reduced ecological flow at many river cross sections and reduced water environmental carrying capacity. Besides, the increase in pollutants caused by the social development makes rivers faced with more pollution pressure. The prominent problems of insufficient water resources and water environmental carrying capacity restrict the sustainable economic and social development and the water eco-civilization in the basin.

The integrated basin water environment treatment is an important measure for implementing the ecological modernization and plays an important role in promoting the transformation development and green development of Chongqing. According to the requirements for eco-civilization and green development, Chongqing shall strengthen the integrated water environment treatment, moderately raise the water environment treatment standards and promote the transformation and upgrading of the industrial structure by raising environmental standards. Meanwhile, the strategy of main functional areas shall be implemented and the protection of key ecological areas shall be strengthened to coordinately complete the integrated water environment treatment in basin by taking multiple measures simultaneously.

The causes of river water eco-environmental problems are complicated, involving the river banks, underwater, upstream, downstream and other geographical basin areas and also involving administrative departments for water resources, environmental protection, housing and development departments, etc. The traditional water environment treatment has the problems of localization, fragmentation and linearization, lacking systematic treatment idea, lacking trans-departmental coordination, lacking basin scale-related thinking and lacking integrated water quality and quantity thinking. Moreover, the single measure makes it difficult to achieve precision treatment. The treatment effect is always limited and short term, leading to repeated management but repeated failures in really improving the basin water environment. In order to systematically optimize the water eco-environment treatment, we shall, from the perspective of the basin, deal with the generation and transport of pollutants discharged into river and their impacts on the water environment, and according to the water control ideas of "giving priority to water saving, spatial balance, systematic treatment and joint efforts of government functions and market mechanism" in the new era, insist on "promoting well-coordinated environmental conservation and avoiding excessive development"; strictly adhere to the basin ecological protection red line centered on water quantity, water quality and water ecological elements; and propose the trans-departmental and trans-disciplinary basin water environment management mode.

The application of sponge city construction concept to the urban development process has greatly improved the urban stormwater management system and the water quality of urban water bodies. The construction of sponge city mainly includes two aspects: the first is to build "sponge bodies" used for stormwater runoff storage, including rivers, lakes, ponds and other water systems, as well as urban supporting facilities, such as greenbelts, gardens and permeable roads; another aspect is to sort out the urban storm and sewage networks in the urban drainage system, scientifically construct stormwater lift stations to enable stormwater to be drained through the pipe networks and pump stations, so as to increase the water discharge, effectively improve the urban drainage system standards and reduce the pressure of urban waterlogging.

With the increasing urban development intensity, continuously disappeared green facilities and decreased green space area have led to the decline in natural retention of stormwater and damage to water eco-recycling, resulting in adverse impacts on the regional ecology, water environment and hydrogeology. Meanwhile, China is short of water resources, with its total water resources only accounting for 6% of the global water resources, while the comprehensive utilization of collected stormwater is one of the important means to alleviate the water shortage situation. The sponge city is a positive urban construction concept that should be implemented and popularized.

1.4.2 Significance of integrated basin water environment treatment

1.4.2.1 Social benefits

It can ensure the production and life safety of residents. A large amount of raw sewage is directly discharged into water bodies and soil, which may cause the ecosystem destruction in the water bodies and soil or ecological unbalance. If the soil is polluted by heavy metals and synthetic organics, the content of heavy metals and other hazardous substances in agricultural products may badly exceed the standards, directly affecting the human health. Therefore, the water treatment is one of the means to ensure the safety of human production and life.

It can improve the ecological environment quality. After the water ecological environment is polluted by phosphorus and nitrogen, the water eutrophication may occur, causing mass propagation of algae, algal blooms, stinking water, deterioration of water quality and other environmental problems. The water resources treatment through reasonable and effective measures can improve the water environment, make the water clean and improve the quality of the ecological environment.

It can realize the sustainable development. The water resources treatment engineering such as the reclamation and reuse of sewage is an effective and sustainable method to solve (or alleviate) the problems of urban water shortage and water ecological imbalance of rivers, an effective measure for realizing the Chongqing environmental goal of "leading in west China and first-class across China" and also an effective attempt to actively explore the new way to the construction of a resource-conserving and environment-friendly society.

1.4.2.2 Economic benefits

Reduces the loss of water pollution to agricultural products. Water environment pollution causes various economic losses, mainly reflected in the aspects of agriculture, physical health and ecological environment. The harm of water environment pollution to agricultural products is mainly reflected in the decline in agricultural product quality and output.

Improves the urban competitiveness. With the progress in urbanization process and the need of constantly improving the quality of life in small towns, on the one hand, a great deal of small- and medium-sized sewage treatment facilities adapting to the size of small towns are emerging, so there will be stronger market demand. On the other hand, due to the increasingly precious water resources, the mode combining centralized water recycling treatment system with local small-scale water recycling treatment system is more widely used.

1.4.2.3 Ecological benefits of water resources

Contributes to the regional water environment protection and ecological restoration. The regional ecosystem is increasingly fragile due to great changes in the water environment, while the water resources treatment involves the sewage treatment of large area and is conducive to the regional ecosystem protection.

Promotes the recycling of water resources. The systematic water recycling mode is to establish a water recycling network connecting the agricultural water, industrial water, urban miscellaneous water and ecological landscape water by taking a city as a whole, thereby to realize the recycling of urban system. At this level, industries and enterprises further expand and extend, forming a more complex closed recycling network of water resources in urban areas. The recycled municipal sewage is a potentially available water resource and is gradually becoming a necessary second water source for cities, which can be used for industrial cooling, agricultural irrigation, urban miscellaneous purposes, landscape watering, water source replenishing, etc.

Contributes to increase biodiversity in the water environment. The water resources treatment involves measures such as the construction of ecological parks, constructed wetlands and green buffers, which can protect and increase the biodiversity, improve the potential co-existence of species, promote the protection and restoration of urban biodiversity and provide natural and ecologically sound open spaces for the public. The water resources treatment not only can alleviate the pollution of water bodies but also can promote the development of aquatic biological diversity.

1.4.3 Characteristics of integrated treatment of mountainous basin

Under the background of the new era, the ecological civilization construction needs to be vigorously promoted, the key to which is to realize the sustainable development of the basin water environment. By the overall planning of major economic and social activities for the basin, city, people and water, measures such as integrated water environment treatment, efficient water utilization and water ecological protection and restoration shall be systematically promoted to explore and form the treatment and development patterns suitable for different types of basins.

As the largest mountainous city in the Three Gorges Reservoir Region and with its mountainous area reaching 76%, Chongqing has typical characteristics of a mountainous city: numerous mountains and hills, complicated landforms, rugged terrain, large elevation differences, widely distributed rivers, densely scattered lakes and strong erosion effect. The special natural geographical environment (mountainous region), particular stage of development (under-developed) and key geographical location (Three Gorges Reservoir Region, upper reaches of Yangtze River) lead to that the basin water environment treatment in Chongqing is facing greater challenges. Meanwhile, the mountainous sponge city construction shall, on the premise of ensuring the safety of urban drainage and waterlogging control, realize the retention, storage, infiltration and purification of stormwater in the urban area to the greatest extent, so as to promote the stormwater recycling and eco-environmental protection. The construction of sponge city that organically combines the permeable pavement, green roof and other green facilities with the urban pipe network construction can solve the problem of stormwater discharge from the source and effectively solve the urban waterlogging problem. In addition, the purified stormwater treated with water purification technology can be used for various non-drinking purposes to solve the problem of water resource shortage. Furthermore, various sponge measures can be adopted to comprehensively improve the urban regulation and storage capacity and finally comprehensively improve the urban residential environment. The sponge city construction shall follow the principle of "ecology in priority", combine natural approaches with artificial measures and on the premise of ensuring the safety of urban drainage and waterlogging control, realize the retention, storage, infiltration and purification of stormwater in the urban area to the greatest extent, so as to promote the stormwater resource utilization and eco-environmental protection. The construction of "sponge city" does not mean start all over again to replace the traditional drainage system. Instead, it is a kind of "load reduction" of and supplement to the traditional drainage system to make the most of the role of a city itself. During the sponge city construction, the natural precipitation, surface water and groundwater systems shall be taken into overall consideration to coordinate the water supply, drainage and other water recycling links, and also consider the complexity and long-term performance.

Under the background of Chongqing's positioning of "two points", "two lands" and "two highs" as put forward by China's General Secretary Xi Jinping and the Yangtze River Protection Program, the core and key point for Chongqing's construction of "a place with beautiful mountains and rivers" is how to protect the basin water ecology, treat the basin water environment, efficiently utilize the basin water resources and embody the development concept of "ecological civilization". According to national ecological civilization construction requirements and current water environment treatment situation in Chongqing, this project intends to put forward a strategic idea for the integrated basin water environment treatment in Chongqing and explore the practical experience in the integrated basin water environment treatment of mountainous city by innovating system and mechanism, aiming to promote China's integrated basin water environment treatment and the strategic goal of sustainable development in a point (Chongqing) to area (China) manner and finally promoting the basin ecological civilization construction across China.

Chapter 2

Situation before integrated water environment treatment in Chongqing basin

2.1 Overview of the basin in Three Gorges Reservoir Region

2.1.1 Distribution of water resources in the basin

Chongqing is located at the junction of Yangtze River and Jialing River, about 450 km wide from north to south, 470 km long from east to west, covering an area of 8.24 km². All the rivers in Chongqing belong to the Yangtze River basin, crossing the land forming a radial but asymmetrical network distribution system. Among them, there are 374 rivers with a basin area more than 50 km², 36 rivers with a basin area more than 1,000 km² and 18 rivers with a basin area more than 3,000 km². Dozens of main rivers flowing into Chongqing Three Gorges Reservoir Region are Yangtze River, Jialing River, Wujiang River, Qijiang River, etc., while the main river flowing out is Yangtze River. The main stream of Yangtze River flows across Chongqing from west to east, about 665 km long, and runs across three anticlines of Wushan Mountain, forming the famous Qutang Gorge, Wuxia Gorge and Xiling Gorge (in Hubei Province), namely the well-known Three Gorges of Yangtze River; Jialing River flows from the northwest and merges into the Yangtze River in Yuzhong District after three bends, forming Libi Gorge, Wentang Gorge and Guanyin Gorge, namely the Mini Three Gorges of Jialing River, while Wujiang River mergers into the Yangtze River in Fuling District. Chongqing is the largest inland port city in the upper reaches of the Yangtze River and even in west China, and is also the shipping center in the upper reaches of the Yangtze River. The Yangtze River, Jialing River and their tributaries in Chongqing form the water transport network of the upper reaches of the Yangtze river with Chongqing urban area as the center, connecting 136 navigable rivers. Ports like Wanzhou and Fuling and dozens of passenger and cargo wharfs are established along the rivers, supporting the navigation of 1,000 (10,000)-ton ships all the year round. From 1926 to 1934, there were 40 ferry wharfs along the two rivers; in 1935, the Jiangbei, Qiansimen, Taipingmen, Feijiba, Jinzimen and Chuqimen wharfs were built successively.

Chongqing is rich in transit water resources, with multi-year average transit water volume of 300–400 billion m³, annual average runoff of 9,775.33 m³/s, multi-year average runoff of 51.14 billion m³, theoretical reserves of 13.38 million KW and exploitable capacity of 7.6 million KW. The surface runoff in Chongqing Three Gorges Reservoir Region is completely replenished by the atmospheric precipitation, so its inter-annual variation and intra-annual distribution are positively correlated with the precipitation in the same period, and there are great differences among regions affected by the topographic and geomorphic factors. According to the statistical data over the years, most areas of the Reservoir Region have an annual precipitation of 1,000–1,200 mm, multi-year average runoff depth of 620.7 mm, multi-year average surface water resources of 56.77 billion m^3 and per capita quantity of surface water resources of 544 m^3. The advanced economic circle has multi-year average surface water resources of 2.98 billion m^3 and per capita quantity of surface water resources of 544 m^3; west Chongqing has multi-year average surface water resources of 9.08 billion m^3 and per capita quantity of surface water resources of 895 m^3; the Eco-economic Zone of the Three Gorges Reservoir Region has multi-year average surface water resources of 44.71 billion m^3 and per capita quantity of surface water resources of 2,817 m^3. Besides, the groundwater in Chongqing is also replenished by the atmospheric precipitation and restricted by the geological structure, landforms and water content spatial distribution. Due to the complicated geological structure, the hydrogeological environment is also very complicated. The groundwater in the whole region mainly includes three types, namely carbonatite karst water, clasolite pore-fissure water and bedrock fissure water. According to rough statistics, the annual average total amount of groundwater in the region is about 8 billion m^3.

The characteristics of water resources in Chongqing Three Gorges Reservoir Region are: (1) Uneven distribution of precipitation. The precipitation is abundant in central and eastern areas, but less in other areas. The precipitation is mainly concentrated in northeast and southeast Chongqing, showing a trend of gradual increase from west to east. The hills and flatlands have large population but less water and small area, while high mountains and high hills have small population, less water but large area, with thin soil moisture layer and poor water-retaining capacity. (2) Uneven distribution of surface water resources and abundant transit water resources. The total surface runoff volume is more than 28 billion m^3, but the regional distribution is uneven. The areas of middle and low mountains in central and eastern Chongqing have less population, small area and high water yield; the areas of low mountains and hills in central and eastern Chongqing have large population, vast area and low water yield. There are abundant transit water resources, but the local water resources are scarce, characterized by uneven regional, spatial and temporal distribution and not matching with the productivity development. The rainfall is rich in high mountain area and relatively less in hilly area. (3) Due to its subtropical monsoon climate, Chongqing is confronted with the problem of uneven spatial and temporal distribution of water resources. The intra-annual precipitation distribution is uneven that the precipitation in summer and autumn is higher than that in winter and spring, and the precipitation is mainly concentrated in the period from April to October, accounting for 84.5% of the total annual precipitation.

2.1.2 Analysis on water quality in the basin

After the water impoundment in the Three Gorges Reservoir Region, the main stream flows more slowly in the Reservoir Region, resulting in reduced self-purification capacity and algal blooms in some tributaries.

In November 2011, the water quality of the "three rivers" (the Yangtze River, Jialing River and Wujiang River) was excellent based on nine assessment indexes; the proportions of cross sections with Class I and Class II water quality were, respectively,

Table 2.1 Assessment results of water quality in Chongqing segment of Yangtze River in November 2011

River name	District name	Cross-section name	Cross-section nature	Nine indicator assessment	
				Water quality class	Cross-section status
Chongqing segment of Yangtze River	Yongchuan District	Zhutuo	Sichuan–Chongqing border, state-controlled	II	Excellent
	Jiangjin District	Jiangjin Bridge	Municipality-controlled	II	Excellent
	Dadukou	Fengshouba	Municipality-controlled	II	Excellent
	Jiulongpo	Heshangshan	Municipality-controlled	I	Excellent
	Nan'an District	Cuntan	State-controlled	II	Excellent
	Jiangbei District	Yuzui	Municipality-controlled	I	Excellent
	Changshou District	Shantuo	Municipality-controlled	II	Excellent
	Fuling District	Yazuishi	Municipality-controlled	II	Excellent
		Qingxichang	State-controlled	I	Excellent
	Fengdu County	Daqiao	Municipality-controlled	I	Excellent
	Zhongxian County	Sujia	Municipality-controlled	II	Excellent
	Wanzhou District	Shaiwangba	State-controlled	II	Excellent
	Yunyang County	Kucaotuo	Municipality-controlled	II	Excellent
	Fengjie County	Baidicheng	Municipality-controlled	II	Excellent
	Wushan County	Peishi	Chongqing–Hubei border, state-controlled	II	Excellent

26.7% and 73.3% in 15 cross sections of the Yangtze River, among which the water quality at Shaiwangba cross section in Wanzhou District was excellent, belonging to Class II. Water quality of the Yangtze River main stream in the Reservoir Region was classified as Classes II–IV, and water quality of tributaries was classified as Classes II–V. Table 2.1 shows the assessment results of the water quality at 15 cross sections in Chongqing segment of the Yangtze River.

There were great temporal and spatial variations of water resources in the Yangtze River basin and also great differences in the water quality between the withered and high water periods. The proportions of river lengths of corresponding water quality classes in the total assessed river length are as shown in Table 2.2:

As for overall water quality in the Yangtze River basin: water quality of the main stream in withered water period was superior to that in high water period; water quality of tributaries in withered water period was inferior to that in high water period; water quality near the river banks was inferior to that in the middle of the river; and water quality of river segments in urban areas was generally inferior to that in non-urban areas.

Table 2.2 Proportions of river lengths of corresponding water quality classes in total assessed river length

Period	Main stream				Main tributaries			
	Assessed river length (km)	Proportion (%)			Assessed river length (km)	Proportion (%)		
		Classes 1–2	Class 3	Classes 4–5		Classes 1–2	Class 3	Classes 4–5
Withered water period	4,616	80.0	20.0		6,224	69.2	12.4	18.4
	4,347	72.1	27.9		4,347	66.8	14.9	18.4
High water period	4,616	28.7	58.9	12.4	6,224	61.0	20.3	18.7
	4,347	66.6	20.6	12.8	4,347	62.2	29.0	8.8

In terms of existing surface water, water of the three rivers (the Yangtze River, Jialing River and Wujiang River) in Chongqing, especially the Yangtze River, was of good quality. In 2017, the overall water quality of Chongqing segment of the Yangtze River main stream was excellent, and the monitoring results at 15 cross sections showed that the proportion of cross sections with Classes I–III water was 100% and the overall water quality of tributaries of the Yangtze River was good. Chongqing centralized drinking water sources were of good quality, and 100% of the 64 urban centralized drinking water sources met the water quality standard. The water quality of the Yangtze River was unstable in 2016, but maintained at Class II from 2017 to 2019, so the water quality of Chongqing Segment of the Yangtze River was relatively good. Besides, the water quality of Jialing River also has been maintained at Class II for a long time, while the water quality of Wujiang River was not good before due to the excessive total phosphate content but now stably meets the Class III standards. In 2018, the overall surface water environment of Chongqing was good. The water quality of Chongqing segment of the Yangtze River main stream was excellent, and the proportion of cross sections with the water quality reaching or superior to Class III in the 42 cross sections included into the national assessment was 90.5%, 2.4% higher than the national annual goal; the rate of water quality reaching the standard of all urban centralized drinking water sources in Chongqing was 100%, and 48 urban segments with black and odorous water were basically free of black and odorous water. The groundwater quality in nine places remained stable and no major and serious water pollution incident occurred in Chongqing Segment of the Yangtze River. Among them, the water quality of rivers such as Qijiang River, Renshi River, Longxi River and Daxi River was stably improved to Class III from Class IV, and the water quality of Linjiang River was significantly improved. From January to November 2019, 93.4% of the 211 cross sections in Chongqing met the water functional requirements, up 7.6% from a year earlier. Among them, the proportion of cross sections with the water quality reaching or superior to Class III in the 42 cross sections included in the national assessment was 97.6%, up 7.1% from a year earlier. In 2020, it is planned to achieve that the water quality of 95.2% of the 42 cross sections included in the national assessment reaches or is superior to Class III; realize no cross section in Chongqing with the water quality inferior to Class V, 100% urban centralized drinking water sources with the water quality reaching the standard and more than 86% rural centralized drinking water sources with the water quality reaching the standard.

2.1.3 Current situation of water resources utilization

The water resources of the Three Gorges Reservoir Region are characterized by large total amount but small amount per capita, and there is difficulty in water utilization in some areas and regions. The total amount of water resources in the Three Gorges Reservoir Region is 462.442 billion m^3, while the local water resources per capita and water resources per mu (1 mu = 0.0667 ha) are about 64% of the national averages. According to the data, over the years, the proportion of areas with sufficient rural production water and basic domestic water was 38%, and the proportions of areas short of and desperately short of production water were 37% and 25%, respectively, namely, more than half of the demand for rural production water was not satisfied; the proportion of areas with sufficient or basically sufficient domestic water was 54.3%, and the proportions of areas short of and desperately short of domestic water were 26.4% and 19.3%, respectively, namely, nearly half of the demand for rural domestic water was not satisfied. Meanwhile, with the increases in urbanization rate and urban population, the development, utilization and protection of water resources are still relatively backward. As a result, the water quantity and quality in some areas can't meet the living needs of urban residents, and the problems are acuter especially in the main urban area located in the fluctuating backwater area at the end of the Three Gorges Reservoir.

Due to the greatly undulating terrain and deep river valleys in the Three Gorges Reservoir Region, the utilization of water resources there is characterized by high difficulty and high cost. According to the data in 2005, Chongqing utilized 3.99 billion m^3 of local surface water, accounting for 7.8% of its total water resources, and utilized 160 million m^3 of groundwater resources, only accounting for 1.5% of its total groundwater resources. Therefore, Chongqing had a low utilization rate of water resources and was highly dependent on water conservancy projects, belonging to the engineering-caused water shortage area that also had the problems of pollution-induced water shortage and water resource shortage. Over the years, 189,000 various water supply projects have been built in Chongqing, with an annual water supply of 5.541 billion m^3; and a total of 183,416 farmland water conservancy projects have been built in Chongqing, with the effective irrigation area of 9.45 million mu, solving the drinking water problem of 5.9 million rural people.

2.2 Water environment problems and impacts in the basin

2.2.1 Water ecology problem

2.2.1.1 Current situation of ecological water

Water is not only an important material basis for human survival and development and an essential element for the growth of plants and animals but also an important part of the environment. Ecological water refers to the existing water resources, including surface water, groundwater and soil water, etc., which are necessary to be consumed for maintaining the normal development and relative stability of various ecosystems within a specific space and time but not used as social and economic water. In a broad sense, the ecological water refers to the water consumed for maintaining the ecosystem integrity, which includes a part of water resources and a part of water that is often not included in the water resources calculation, such as the amount of ineffective evaporation and plant interception; in a narrow sense, the ecological water refers to the total amount of water resources necessary for maintaining the ecosystem integrity.

2.2.1.2 Ecological destruction problem

The ecological destruction problem is mainly reflected in the biodiversity reduction and water and soil loss. The biodiversity is an important indicator of ecological balance and an important content of ecological construction. According to statistics, there are about 6,500 species of animal and plant resources in the Three Gorges Reservoir Region, accounting for about 20% of the total animal and plant species in China. Among them, there are, respectively, eight and six species of national Class I protected animals and plants and, respectively, 35 and 22 species of Class II protected animals and plants. Besides, there are also abundant tertiary relic plants in the Three Gorges Reservoir Region. With high mountains, steep slopes, complicated geological structures, deep ravines and densely distributed water systems, Chongqing is one of the areas with the most serious geological hazards and also one of the cities with the most serious water and soil loss in the upper reaches of the Yangtze River, while Chongqing Three Gorges Reservoir Region is the area with the most serious water and soil loss in Chongqing and also the national key ecological function area for water and soil conservation and key water and soil loss treatment area.

2.2.1.3 Function zoning of water body

The function zoning of water body refers to the process of dividing the water body into different zones with the water quality meeting different requirements according to the environmental conditions, use conditions and needs of social and economic development, which is the basic work for the water environment management and water resources development and utilization. After the function zoning of water body, the water pollution control objectives of each zone can be defined, and the key points of protection can be highlighted, which are of great significance for the water source protection, environmental improvement and the promotion of economic development. According to the *Surface Water Environmental Quality Standards*, the water bodies can be classified into seven types by function zoning: ① natural reserve and source water (Class I as per *Surface Water Environmental Quality Standards*); ② drinking water source area (Class II as per *Surface Water Environmental Quality Standards*); ③ aquaculture area (Class II as per *Surface Water Environmental Quality Standards*); ④ tourist area (Class III as per *Surface Water Environmental Quality Standards*); ⑤ industrial water area (Class IV as per *Surface Water Environmental Quality Standards*); ⑥ agricultural irrigation water area (Class V as per *Surface Water Environmental Quality Standards*); and ⑦ mixed area (zone) near sewage outfalls (*Surface Water Environmental Quality Standards* not applied).

According to relevant data analysis, among the economic losses of various water body functions in the Reservoir Region, the industrial function suffered the greatest economic loss amount, followed by the economic loss of agricultural function, while the economic loss amount of domestic water function and that of the fishery function were relatively small. The economic loss amount of each function of the water body was related to the output value of the corresponding industry, and the change trend and amplitude were basically consistent with the overall functional economic loss.

2.2.2 Water pollution problem

2.2.2.1 Situation of water pollution discharge

With the acceleration of urbanization and industrialization processes, the discharge of sewage in the Three Gorges Reservoir Region also increases. Domestic waste indirectly enters water bodies in the Three Gorges Reservoir Region under the effects of surface runoff and stormwater, polluting the water environment. Besides, the pollution problems caused by pesticides, fertilizers and concentrated aquaculture also exist in the Reservoir Region.

2.2.2.2 Surface water pollution problem

As for the surface water pollution problem, the pollutants produced by human activities have a series of physical and chemical effects on the water and sediment in the surface water, leading to changes in the physical and chemical properties and changes in the biocenosis composition, resulting in surface water quality pollution and affecting the functions and use of the water. The surface water pollution mainly comes from industrial wastewater and domestic sewage. Due to the complex composition, the sewage and wastewater contain a variety of organic and inorganic pollutants. Besides, the hospital sewage also contains a large number of pathogens.

The pollution of different water bodies has different characteristics, which can be divided into three types according to the types of surface water bodies, namely river pollution, lake (reservoir) pollution and marine pollution. As for river pollution, the degree of water pollution varies with the runoff volume and the discharge quantity and manner of sewage: the fast-spreading pollutants will cause greater pollution impact; the upstream pollution can quickly affect the downstream with the flow; the pollution in a channel segment will affect the aquatic biological environment of the entire channel; and pollutants in rivers can harm the human through drinking water, farmland irrigation and food chains. The lake (reservoir) pollution is characterized by the long-term retention of some pollutants, which will lead to the quantitative accumulation and qualitative changes, primarily the lake eutrophication caused by phosphorus, nitrogen and other plant nutrients. The marine pollution is characterized by diverse and complex sources, long persistence, severe harm and wide affected range. In addition to the pollution from ships and offshore oil wells, the industrial wastewater and urban sewage discharged in coastal and inland areas flow into seas eventually, harming marine organisms and destroying marine resources. Therefore, the marine pollution treatment has become an important aspect of water environmental protection.

The surface water has diverse pollution sources, mainly from the direct discharge and leakage of toxic and harmful wastewater or sewage, and improperly disposed waste flowing into water sources due to precipitation, flash floods and other reasons, specifically as follows:

1. Sewage discharged by nearby residents and factories, namely domestic sewage and industrial wastewater;
2. Waste discharged from nearby wharfs and warehouses that are inappropriately planned and used;
3. Domestic sewage and waste discharged from ships;
4. Street dust, rubbish and other pollutants on the ground, as well as fertilizers and pesticides applied to farmland carried by surface runoff into water;
5. Washing dirt and impregnating industrial materials in water.

2.2.2.3 Groundwater pollution problem

Groundwater resources not only are significant in quantity but also have the advantages of good water quality, wide distribution and convenient in situ exploitation and utilization, which are important resources for the human survival and development. The groundwater in Chongqing is composed of geothermal water, carbonatite karst water, clasolite fissure water and red bed pore-fissure water, with an annual reserve of 13.17 billion m^3 and exploitable amount of 4.49 billion m^3. The exposed area of carbonate rocks is 2,903 km^2, accounting for 35.3% of the total area of Chongqing. The karst water accounts for 78% of the total amount of groundwater, mainly distributed in areas of Daba Mountain and Wuling Mountain, while the bedrock fissure water only accounts for 6%, distributed in the red bed hilly area in west Chongqing.

After being tested, the geothermal water is of good quality, with normal water quality indicators, while the red bed pore-fissure water has been of relatively poor quality over the years, with indicators such as Fe, Mn^{2+}, COD_{Mn}, bacteria, *Escherichia coli* and NH_3-N exceeding the standards and other indicators within the normal range, which is related to the primary geological conditions and man-made pollution. The carbonatite karst water is of poor quality, with its six water quality indicators, namely Fe, COD_{Mn}, *E. coli*, NH_3-N, total bacterial count and F^- once exceeding the standards and other indicators within the normal range, which is closely related to man-made pollution. The clasolite fissure water has also been of relatively poor quality over the years, with three indicators, namely Fe, Mn^{2+} and COD_{Mn} once exceeding the standards, mainly due to the relatively high content of Fe and Mn^{2+} in the primary geological environment.

The precipitation with a pH value less than 5.6 is called acid rain. For many years, the pH value of precipitation and pH value of acid rain in Chongqing main urban area have been below 5.0. The southwest region centered by Chongqing and Guiyang is the second largest acid rain region in China, second only to the central China acid rain region. Although the frequency and acidity of acid rain have decreased gradually in recent years, the acid rain pollution remains a serious problem. The acid rain in Chongqing belongs to sulfuric acid rain, in which the major pollution component is SO_4^{2-} and the major cations are NH_4^+ and Ca^{2+}. Nitrate is one of the most important pollutants in groundwater, so nitrate nitrogen is taken as the key indicator to assess the groundwater quality at home and abroad. The sources of nitrogen pollution in groundwater mainly include the use of nitrogen fertilizer, industrial wastewater, domestic sewage, acid rain, garbage accumulation, human and animal excreta, etc. The degradation and nitrification of natural organic nitrogen or humus are potential sources of the nitrate in groundwater. Among them, the pollution caused by the massive use of nitrogen fertilizer in karst areas is particularly serious in the vast rural areas, which is worthy of in-depth research and discussion. Due to the fracture development in karst areas, nitrogen fertilizer, excreta, garbage and other nitrogenous substances can easily be carried by stormwater to the groundwater through soil, karst channels and ponors, making the nitrogen-containing compounds converted into nitrite nitrogen through the aforesaid various channels, animal and plant residues and the fixation of atmospheric nitrogen, and finally oxidized to nitrate nitrogen by nitrobacteria. Except for the part absorbed by plants, most of the remaining nitrate nitrogen seeps or flows into the groundwater system, causing groundwater pollution.

In recent years, the groundwater pollution has been prominently reflected in many aspects, such as the waste landfilling and leaching, the pesticides and fertilizers seeping into groundwater along with stormwater, the petroleum and chemical products, benzene and its homologs, phenol, high-molecular polymers and other refractory organics seeping into the aquifer, which have seriously polluted the groundwater and caused serious harm to human beings.

2.2.2.4 Black and odorous water problem

Water resource is an indispensable and important resource for the human life and work. However, due to the increasingly accelerating urban development in China, the rivers in some cities are seriously black and odorous. The black and odorous water body refers to the phenomenon produced when the water body is polluted by organic pollutants exceeding its self-purification capacity, characterized by obvious black color and stinking odor, basically no plankton in the water, serious degradation of aquatic plants, fracture of food chain, ecological imbalance of water body and basic loss of function. The black and odorous water body has a significant impact on the urban ecology and a big impact on the daily life of urban residents. The serious phenomenon of black and odorous water body will pollute the water quality, stink and affect the air quality.

The main impacts of black and odorous water bodies on the environment are as follows:

There are great changes in both the water quality of water bodies and the physical and chemical properties of geology, leading to the loss of basic functions of water bodies. Water pollution will cause the water eutrophication and dramatic increase in bacteria, which will lead to the consumption of a large amount of oxygen in the water, directly causing the death of some life in the water. If such situation is not improved for a long time, it will lead to the aging or even death of water bodies.

The black and odorous water can release toxic substances, which will gradually pollute the air, affect the lives of nearby residents and make it hard to ensure the health of residents.

The long-term existence of black and odorous water will reduce the ecological environment quality of the whole city. If not effectively treated for a long time, it will eventually cause a serious impact on the ecological balance of the whole city through continuous accumulation and infiltration, resulting in incalculable harm.

Due to the limited self-purification capacity of some water bodies in Chongqing and the impacts of point, non-point source and endogenous pollution, some water bodies become black and odorous sometimes. There are many causes of black and odorous water, mainly including the following factors:

①   Hydrodynamic conditions.
As one of the important factors in water bodies, the problems caused by hydrodynamic conditions often lead to the black and odorous water, mainly because the obstructed water flow can directly affect the water sample and pollute the water body to different degrees, and the water body will become black and odorous when the water quality is deteriorated.

②   Pollution of metallic elements.
When there are a lot of substances like iron and manganese elements in the water, the reduction reactions will occur under the circumstance of insufficient oxygen that the iron and manganese will react with sulfur to generate substances like ferrous sulfide and manganese sulfide, resulting in black and odorous river water.

③   Organic pollution.
The root cause of the black and odorous water is a kind of biochemical reaction. In the event of oxidative decomposition, the oxygen consumption rate of organic matters in water is obviously higher than the reoxygenation rate, which aggravates the anoxic phenomenon in water body. Besides, the anaerobes in water will produce malodorous pollutants like hydrogen sulfide, methane and ammonia gas in the event of decomposition, which will lead to the obvious phenomenon of black and odorous water with the lapse of time.

In recent years, the urban black and odorous water has been a water environment problem of great concern. The treatment of urban black and odorous water is also one of the core contents of water treatment. The treatment of urban black and odorous water bodies is a systematic task that involves a wide range of aspects. As specified in the *Water Pollution Control Action Plan* ("Ten Measures for Water Pollution Control") published by the State Council, the people's governments of cities shall be the responsibility subjects for the treatment of black and odorous water bodies, and the Ministry of Housing and Urban-Rural Development (MOHURD) shall take the lead to guide the local implementation and propose goals together with ministries and commissions such as the Ministry of Environmental Protection, Ministry of Water Resources and Ministry of Agriculture: by the end of 2017, cities at the prefecture level or above should realize no large area of flotsam on river surface, no waste on riverbanks and no illegal sewage outfall, and the black and odorous water bodies would be basically eliminated in the built-up areas of municipalities directly under the central government, provincial capitals and cities specifically designated in the state plan; by the end of 2020, the black and odorous water bodies would be controlled within 10% in the built-up areas of cities at the prefecture level or above; by 2030, the black and odorous water bodies would be generally eliminated in the built-up areas of all cities in China. In 2015, 31 segments of black and odorous water were found in Chongqing main urban area, with a total length of 143.6 km. After more than 2 years of pollutant interception, river endogenous pollution source treatment and ecological restoration, the water was basically free from "black" and "odorous". According to the result of national special inspection in June 2018, Chongqing became one of the nine cities in China that had a 100% elimination rate of black and odorous water bodies. Since 2015, Chongqing has promoted the treatment of black and odorous water bodies as a project for the people's livelihood. It firstly carried out the integrated treatment of 56 lakes and reservoirs, then made up for its weaknesses, such as in the aspect of domestic sewage pipe network. Therefore, in the special inspection organized by the MOHURD and Ministry of Environmental Protection in 2018, Chongqing was one of the first-batch cities that eliminated black and odorous water bodies. The water quality of main streams in Chongqing is relatively good, but that of some tributaries still does not meet the standards, so we need to make joint efforts to keep improving the water quality.

2.2.3 Water safety problems

2.2.3.1 Water supply safety

Water supply safety—all water demands inside and outside buildings are satisfied, and there is sufficient supply of water with guaranteed quality and appropriate water pressure available at any time for places in the need of water.

Water supply system consists of a series of water treatment buildings and water transport and distribution pipelines, which is the combination of the facilities for the water source intaking and transport, sedimentation and filtration treatment and distribution. Its main tasks are as follows: water supply works take water from water resources and process and treat the water according to the requirements of water quality standards, then transport the water to water-consuming areas in cities and towns through the pipelines to distribute water to users. The work procedure of water supply engineering system usually includes three parts: water intake engineering system, water treatment engineering system and water transport and distribution engineering system. The water intake engineering system consists of intake structures and primary pump stations for transporting water from the water intake to water works. Its main function is to ensure that sufficient water is obtained from the water source and transported to the water works. The water treatment engineering system mainly consists of various water treatment equipment. Its main function is to make the water quality meet the national quality standards for drinking water and production water through the water quality standard treatment process. The water transport and distribution engineering system mainly consists of secondary pump stations and water transport and distribution engineering system pipelines. Its main function is to transport treated water to various water-consuming ends in the cities and towns, so as to ensure adequate water pressure in the water supply system.

The urban water shortage in China is mainly due to the shortage of water resources, for that the per capita water resource in China only accounts for 22% of the global per capita water resource. China is listed in the countries with water shortage. The imbalance between the water supply and demand is particularly prominent in China's areas with developed economy and large population, especially south China with serious water resources pollution. With the acceleration of urbanization and industrialization processes, the scale and water quality of existing water supply facilities are more difficult to meet the development needs. The engineering-caused water shortage is prominent in Chongqing. Considering the impact of water source quality on regional water supply safety and the phenomenon of serious water supply area segmentation, the water supply safety and reliability become more important. The water supply safety usually has the following problems:

1. Water source pollution. The water pollution has still been exacerbated in recent years. Rivers and lakes in and around cities and towns have been generally polluted and the water pollution is especially serious in industrially and economically developed areas, further aggravating the water crisis in cities and towns.
2. Serious aging of water supply equipment. As the water supply systems in some cities and towns were designed and constructed early, the equipment is very old and backward and the groundwater resource is generally used as the water supply source, which is hard to meet the living and production demands of the whole city or town. Many enterprises have built their own water intake and supply equipment, but these water supply pipelines are not connected to each other, resulting in repeated construction of water supply equipment, low use efficiency and waste of resources, which is not conducive to the rational utilization of water resources. During the peak period of water use in summer, it is hard to meet the water demands and guarantee water supply safety even if the water works are operating at full capacity. First of all, the water source safety can't be ensured due to the large water consumption volume in summer. What's more, the production of water works is affected by the power supply in summer. The power cut will lead to water outage, which is the largest threat to the safe production of water works. In addition, it is hard for water works to ensure the water supply volume in withered water period.
3. The construction of urban water supply network construction cannot keep pace with the development of times. Most of the urban pipe networks were planned, designed and constructed according to the water supply volume for the populations at the time of construction, during which the per capita water consumption was lower. With the continuous development and construction of cities and towns, the water consumption has increased sharply. Although the water supply networks have been continuously extended and expanded, the pipe diameters have already been unable to meet the future development needs of the cities and towns, and the water supply networks are often overloaded. Factors such as the breakage and leakage of pipes, small diameters of pipes designed and constructed and pipe scaling may also cause the problem of insufficient water supply pressure. Besides, the fund shortage is also a main reason why the construction of water supply pipeline lags far behind the development and construction of water works, which causes the water transport capacity of the water supply pipeline not matched with the production capacity of water works, resulting in the production capacity waste of water works and then the imbalance of water supply demand.
4. The urban water supply is mainly for domestic use, but the proportion of industrial water is increasing year by year. The urban water supply engineering is mainly for supplying domestic water to urban residents, including the domestic water and commercial economic water of urban residents, while the water for urban industrial production is generally supplied by self-built facilities. With the rapid

development of urban economy, the consumption of urban industrial production water and domestic water is increasing year by year. Especially in economically developed areas in south China, the urban industrial water consumption is increasing rapidly. The construction of urban water supply engineering system provides convenience for the rapid development of industrial production and accelerates the increase in the proportion of urban industrial production water, so as to promote the rapid construction of the urban water supply engineering scale and increase economic benefits.

5. Shortage of funds for the construction of urban water supply system. The water supply engineering is huge systematic engineering characterized by large quantities of design and planning, complex construction process and long construction period, so it requires huge fund investment. Especially in the minds of the public, it has been a long-standing idea that water is a social welfare cause, so the tap water is not correctly regarded as a commodity, which makes it impossible to implement the water supply industry in the modern market economy mechanism. As a result, the water supply industry in most of the cities and towns is always loss-making, with the operation and production solely relying on the state and government financial subsidies, resulting in the general shortage of construction funds for urban water supply system engineering and the failure of water supply industry to keep pace with the modern urban economic construction.

2.2.3.2 Water drainage safety

Due to the complex geological and topographical conditions of mountainous city in the Three Gorges Reservoir Region, there are frequent geological disasters such as landslide, debris flow and foundation settlement, usually leading to breakage and damage of drainage pipelines; due to the complex layout and many damages of urban drainage pipelines, the operational safety problems such as blockage and breakage of pipes are serious; the large topographic slope of the mountainous city makes the non-point source pollution more serious, which has an obvious impact on the water quality. With the rapid development of urban construction, the runoff generation and convergence conditions have been changed obviously, making the capacity of existing stormwater drainage systems obviously inadequate. Therefore, to ensure the people's normal life and production, the drainage safety must be taken seriously.

Drainage safety—all the wastewater and sewage inside and outside buildings can be drained out of the buildings and parks smoothly and timely without blockage, overflow or sedimentation; no wastewater or sewage that contains excessive pollutants shall be discharged into municipal pipelines or receiving water bodies; the wastewater or sewage discharged shall not give off an unpleasant odor in the environment; there shall be channels available for the drainage of water for fire-fighting in case of a fire.

Drainage engineering facilities, including waterlogging prevention and control facilities, stormwater management, storage and utilization facilities, are important infrastructure to maintain the normal operation of cities and towns and the resource utilization, which are particularly important in areas with frequent rain, dense river network or vulnerable to waterlogging disasters. The sewage treatment rate of drainage facilities and the centralized treatment rate of sewage treatment plants in

Chongqing are higher than the national average. However, due to the inadequate sewage treatment facilities constructed, relatively lagging sewage network construction and unbalanced regional layout, the sludge treatment scale still needs to be improved and the stormwater collection and discharge systems are still inadequate.

According to the requirements of urban drainage planning, the rain and sewage separation shall be implemented in the existing combined sewer system; as for the sewer system temporarily lacking the conditions for rain and sewage separation, the measures combined with interception, regulation, storage and treatment shall be taken to increase the interception ratio and strengthen the pollution prevention and control during the initial stage of rainfall. According to the current situation of drainage pipe network construction in China and the requirements of urban drainage planning, Chongqing shall accelerate the urban drainage pipe network transformation and implement the rain and sewage separation. Meanwhile, the interception ratio shall be increased and combined measures of interception, regulation, storage and treatment shall be taken to reduce the pollution of combined sewage and initial stormwater.

With the goals of protecting the quality of regional drinking water sources and protecting the water environment of rivers, river systems and reservoirs in basins of the Yangtze River and Jialing River, the construction level of sewage collection and treatment facilities and the urban stormwater discharge capacity shall be improved, so as to make the sewage treatment and stormwater discharge level in areas within the second ring road in the leading position in west China and ensure the overall water environment quality of the Yangtze River, Jialing River and Three Gorges Reservoir Region meets corresponding standards. The construction of sewage collection facilities shall be strengthened, perfect urban sewage collection, transportation, treatment and discharge systems shall be built and the urban sewer system transformation shall be completed to ensure that the sewage discharged meets corresponding standards. The construction of sewage treatment facilities shall be strengthened to make the concentrated treatment rate of urban sewage treatment plants within the second ring road reach 95% and the domestic advanced level by 2020. The industrial wastewater discharged shall meet corresponding standards, with 98% of the industrial wastewater discharged meeting the standards. The sludge of sewage treatment plants shall be properly disposed and subject to resource utilization if possible. The standards for the construction of stormwater collection and drainage systems shall be raised to build stormwater drainage and utilization systems that adapt to the urban development.

2.3 Cause analysis of impacts on basin water environment

2.3.1 Mixing of rain and sewage

At present, the sewer systems of China are mainly divided into three types: separate system, combined system and mixed system (namely part separate system and part combined system). The separate system means that the rain and sewage are transported by two separate drainage pipe networks that don't interfere with each other, while the combined system means that the rain and sewage are transported by the same pipe network. In general, emerging cities are adopted with separate systems, while some old cities and urban areas are still adopted with combined systems and mixed systems due to limited conditions.

The hazards brought by the mixing of rain and sewage are mainly reflected in two aspects: first, the sewage will enter the municipal stormwater pipes and then be discharged into rivers, which will pollute the river water and endanger aquatic plants and animals; second, the stormwater will enter the municipal sewage pipes, which will cause the sewage pipe overflow and potential safety hazards and also has a great impact on sewage treatment plants. When the inflow exceeds the capacity limit of sewage treatment plants, the sewage will be directly discharged, polluting water bodies. Most cities and towns can meet the requirements of "no puddle within the drainage standard and rapid drainage beyond the drainage standard" under normal circumstances, but in case of a heavy rain, the problem of the mixing of rain and sewage will be sharply magnified and a large amount of stormwater will enter the sewage pipes. Due to the discharge capacity limitation of sewage pump stations, a large amount of stormwater will be retained in sewage pipes, which may aggravate the situation of road puddles and water inflow in residential buildings, or even cause safety accidents such as well lid displacement, pipeline breakage and casualties. The local overflow of sewage will flow into nearby rivers through road stormwater collection inlets, aggravating the water pollution.

Main causes of the mixing of rain and sewage are as follows:

2.3.1.1 Mixing of rain and sewage already occurred upon the completion of drainage pipe network construction

Some newly built residential quarters in many cities and towns are designed with separate rain and sewage systems, but the mixing of rain and sewage was already found upon the completion. A large amount of domestic sewage needed to be pumped and discharged at stormwater pump stations every day, indicating that the mixing of rain and sewage in drainage networks of some newly built residential quarters occurred upon the completion because the construction organizations failed in strictly following the design drawings during construction, resulting in disorganized connection of rain and sewage pipes. The mixing of rain and sewage is more serious in some other residential quarters built earlier.

2.3.1.2 Mixing of rain and sewage in municipal pipes

1. Some municipal sewage facilities are not constructed in accordance with the planning, so there is no sewage drainage outlet in some areas and the sewage is discharged into nearby stormwater drainage system.
2. Due to the complexity of urban underground pipe networks and the lacking of expertise on rain and sewage drainage and negligence of contractors and construction workers, the rain and sewage pipes are connected incorrectly. Besides, correct acceptance procedures are not strictly followed after completion, resulting in the mixing of rain and sewage.
3. During the construction of major projects, the normal drainage of nearby rain and sewage is usually affected by temporary drainage measures, namely, temporary measures such as the displacement and inverted siphon setting adopted for the affected rain and sewage pipes during the construction. Due to the unreasonable design of temporary drainage measures, long construction period, slow progress and lax regulation and supervision of projects, these temporary drainage

pipes are often troubled by sediment accumulation and difficulties in maintenance and dredging, so in case of a pipe blockage, there is no way but mixed connection of pipes.

2.3.1.3 Mixing of rain and sewage in some enterprises

In order to reduce the cost and cut down the sewage treatment expenditure, some enterprises directly discharge the untreated sewage into stormwater wells, causing the mixing of rain and sewage.

2.3.1.4 Mixing of rain and sewage caused by weak environmental awareness of residents

Residents lacking of drainage knowledge often only try to ensure the smooth drainage of their own houses to meet their own needs. For example, they may privately transform the balcony into a laundry room, or pour garbage and other waste directly into the sewer that may cause blockage, resulting in the universal phenomenon of the mixing of rain and sewage.

2.3.2 Drainage network construction and sewage treatment

For the sewage treatment, the sewer system shall be selected according to local conditions, and the planning and construction of the sewer system shall be coordinated with the overall planning and construction of the urban sewage treatment plants and sponge cities. For pollution point sources that are directly discharged into water bodies, pollutant interception measures shall be taken and the sewage collection system shall be improved to realize complete collection and complete treatment. Besides, the wrong and missing pipe connection shall be checked.

As for new urban area with abundant rainfall, the separate sewer system shall be implemented, and the overflow combined sewer system shall be gradually transformed to fully combined sewer system and the overflow ports for direct discharge into rivers shall be gradually blocked. As for cities adopted with separate sewer system, if economic conditions permit, the storage facilities such as storage ponds and stormwater tanks will be gradually established to collect, treat and utilize stormwater runoff, so as to control pollution and reasonably utilize stormwater. In arid and semi-arid areas, if there are relatively complete urban sewage pipe networks and sewage treatment plants, the combined system can be adopted to fully utilize pipes and storage facilities to intercept sewage beyond the treatment capacity of sewage treatment plants, so the mixed sewage after rain can be fully treated in sewage treatment plants, thereby reducing the pollution load in rainy season.

In old urban areas, due to the basically fixed underground pipelines, crowded surface structures and narrow roads, the adoption of separate system will cause many realistic questions such as incomplete separation of upstream rain and sewage, huge investment and construction difficulties. Therefore, the selection of sewer system shall not excessively rely on the separate system. Considering the practical factors, most of the old urban areas in many cities will still be in state of coexistence of combined system and separate system for quite a long time.

Most of the old urban areas are adopted with the intercepting combined system, namely overflow control facilities arranged along river banks and lake shores. The original combined pipes are utilized and intercepting pipes are laid along both banks of rivers to collect the stormwater. The specific approach is to set up intercepting wells before main intercepting pipes. The diameter of main intercepting pipe shall be determined according to the dry weather sewage flow ad interception ratio, theoretically making the dry weather sewage and initial stormwater enter the main intercepting pipes. When the increased stormwater exceeds the transport capacity of the main intercepting pipes, the excessive water, including part of sewage and stormwater overflow, will flow into rivers. Such drainage system has achieved certain results in the concrete implementation of sewage interception in old urban areas. It makes full use of the original combined pipes and avoids a large amount of complex and difficult treatment work, so it not only reduces the impacts of sewage pipes laid on urban roads on the road traffic and surrounding residents, saves a lot of investment and solves the problem of initial stormwater pollution but also is easy to be implemented, which may be a relatively good sewage collection system for some areas within a certain period of time.

In addition to further standardizing the combined sewer system, the effect of source control and pollutant interception can be further improved during the implementation through the following ways:

1. Considering the inevitable mixed connection of rain and sewage pipes in separate systems, the dry weather sewage pump can be added in the stormwater pump room when there is a stormwater pump room at the end of pipeline, so the water in dry season can be pumped up into the sewage pipe by sewage pump to avoid the stormwater system's discharge into rivers in dry season.
2. Initial stormwater regulating storage ponds can be established. If it is difficult to transform all the original combined systems into separate systems, the interception ratio can be appropriately increased or the initial stormwater regulating storage ponds can be set up, for example, the stormwater regulating storage ponds set up along the Suzhou River in Shanghai, which will have good effects on reducing the overflow and pollutants discharged into rivers.
3. A sponge city system can be established. As the urban non-point source pollution is mainly from the pollutants contained in stormwater runoff, the stormwater runoff can be effectively controlled by infiltration facilities and retention facilities to reduce the occurrence of combined sewage overflow. Besides, clearing up the garbage around water bodies is also an important measure to control the non-point source pollution.
4. New sewage collection technologies can be actively adopted. For old urban areas where it is hard to collect sewage by traditional gravity pipes, new technologies can be considered, such as the outdoor negative pressure suction sewage collection system. In the said system, the principle of negative pressure suction is applied and one sewage collection well is set up for every one or more households or areas, the bottom of which is connected with a water-sealed suction pipe to form a water seal in the lower part of the water-sealed suction pipe. Driven by the negative pressure in the negative pressure station, the sewage will enter the collection pipe through the water-sealed suction pipe. If this technology is applied, the shallow-buried

pipes can be used to reduce the construction quantities. The practice has proved well on the sewage collection in old urban areas.

5. The monitoring and control of the pipe network operation state can be strengthened, especially the real-time monitoring of the water level, flow rate and overflow water quality of the main pipe intercepting wells, so as to timely know about the operation state of the intercepting wells and to control the initial stormwater intercepting flow through measures such as the intercepting well valve regulation.

Chapter 3

Integrated treatment strategy

3.1 Coordinated treatment of whole basin

3.1.1 Integrated water environment treatment of whole basin

3.1.1.1 Situation of water environment and water ecology in the basin

With the rapid social and economic development of China, the basin water environment quality is threatened and water booms develop frequently, restricting the sustainable social and economic development of China.

3.1.1.2 Basin water circulating process and pollution cause analysis

A basin refers to the specific region formed under the effect of the catchment and movement of water, the surface runoff and river channel are the major characteristics of the basin substance transfer. The main carrier for the pollutant transfer is the water movement and the main way for pollutants transferred from sources to lakes is the basin river system. Therefore, mastering the hydrodynamic characteristics of the basin is the key to the basin water environment treatment and understanding the basin water circulating process, and pollution cause is the basis for the basin water environment treatment.

The basin water circulating process and pollution cause analysis shall be studied as the key basic science problems to find out the production and laws of entering rivers of point source and non-point source nutrients, discuss the transport processes of nutrients in rivers and river systems, reveal the mechanism of nutrient transformation between different interfaces, such as between land area and water area, river and lake, surface and subsurface, and master the effect law of hydrodynamic characteristics on the pollutant transport and transformation in the basin, so as to provide scientific bases for the establishment of a theoretical system for integrated basin water environment treatment with Chinese characteristics and for the guarantee of the ecological environment safety and social and economic development in the basin.

3.1.1.3 Environmental impacts and ecological effects of water conservancy projects

Water conservancy projects have played a significant role in the social and economic development, which have ensured the safety of flood control and drainage, provided the water for life and production and changed the situation of poverty and backwardness and water supply counting on the weather. However, traditional water conservancy projects have indeed made certain negative impacts on the ecology and environment, such as blocking the natural flow of water, weakening comprehensive functions of the ecosystem and worsening the environmental quality of local waters, specifically reflected in the following aspects:

1. River channel straightening projects have sped up the flood flow velocity, increased the flood flow volume, shortened the flood time, improved the flood control safety, safeguarded the life and property and stabilized the social order. At the same time, however, these projects have changed the natural water systems, simplified the ecological structure, reduced the biocenoses, shortened the retention time and weakened the pollution purification capacity and reduced the environmental quality, leading to ecological degradation.
2. River channel hardening projects have reduced the water leakage, improved the water utilization rate, slowed down the slope erosion, maintained the embankment stability and simplified the management of rivers and lakes. At the same time, however, these projects have consumed huge investment, changed the natural system, simplified river functions, occupied waterfront wetlands, blocked the passages between water and land, destroyed the river habitats, weakened the pollution purification capacity, reduced the environmental quality and damaged the landscape structure, leading to ecological degradation.
3. The basin system reservoir (lake) regulation projects have improved the utilization rate of water resources, improved the local climate, ensured the rapid social and economic development, improved the living standard of people and realized the regulation of water between high water period and withered water period. At the same time, however, these projects have reduced the ecological water volume of river base flow, aggravated the river cross-section shrinking, increased the total discharge of sewage, changed the agricultural irrigation and drainage system, increased the proportion of non-point source pollution into rivers, sped up the non-point source pollution into rivers and worsened the water environmental quality of the downstream rivers and lakes.
4. The water gate, dam and station control projects of river systems in the basin have regulated the processes of flood peak and volume, controlled the random flow of water, raised the water level of local waters, improved the irrigation water conditions, increased the retention time of water bodies, inhibited the transport and diffusion of pollutants and prevented the pollutants from transferring to other places. At the same time, however, these projects have intercepted the natural flow of water, blocked the transport of aquatic organisms, accumulated pollutants in water bodies, deteriorated the local water environment quality and increased the risk of water pollution accidents.

The positive and negative effects of water conservancy projects shall be fully and comprehensively recognized, so that the win-win results in the economic and social development and ecological environment protection could be achieved, making water conservancy projects truly and comprehensively serve the survival and development of the mankind.

*3.1.1.4 Measures to be taken by municipal departments for the basin
water environment treatment*

1. Make comprehensive planning for the water environment treatment and reasonably allocate urban water resources. The serious shortage of urban water resources
causes a very common phenomenon of the competition for water resources among
different places. Therefore, it is necessary to look for more water resource channels and reasonably allocate the existing water resources to achieve a balance
among the domestic water, industrial water, landscape water and rivers and lakes.
Moreover, the comprehensive planning for the water environment treatment basin and special plan for integrated water supply and drainage shall be made as
soon as possible based on the water environment capacity. The overall planning
shall be made for the flood control, ecology and landscape, and the overall arrangements shall be made for the construction sequence of infrastructure such
as water supply and drainage networks of key areas, villages and towns. In addition, the annual implementation plans shall be made according to the planning
to adapt economic and social development to the water environment carrying
capacity.
2. Make prospective planning for the urban construction. The urban construction
is a great event characterized by "contributions in the present and benefits in
the future", which shall not only focus on immediate interests or meet immediate needs. The urban water environment treatment shall follow the principle of
"meeting, restoring, improving and expanding" to realize the benign interaction
and harmonious development of the production and life of mankind and the urban environment. The municipal departments shall comprehensively promote the
development of water environment infrastructure and strengthen the construction
and operation of environmental infrastructure.
3. The implementation of innovation-driven strategy shall be comprehensively promoted to strengthen scientific and technological innovation, improve production
technology and resolutely eliminate backward production process, so as to force
enterprises to transform and upgrade with the water control and actively promote
the replacement of traditional industries by advanced industries, replacement of
human by machines, replacement of market by e-commerce and replacement of
land by space. While improving the development speed and expanding the economic aggregate, the municipal departments shall constantly optimize the industrial structure, upgrade the industrial structure and demand, improve quality
while expanding the quantity, transform the economy while developing, vigorously develop the ecological economy and realize the coordinated development of
the primary, secondary and tertiary industries, exploring a "win-win" road for the
ecological environmental protection and economic and social development.
4. Comprehensively promote the integrated treatment of river courses. The organization and leadership shall be strengthened to form a water control system characterized by collaboration between different levels and coordination at the same
level. All departments at different levels shall deepen understanding, make detailed plans, closely cooperate and vigorously implement to ensure the practical
results of water environment treatment. The "river chief system" management
shall be fully implemented and the "one file for one river" shall be established to

record basic information, water quality, sewage outfall, water environment and water ecology of every river, so as to establish the register of river. The "one strategy for one river" shall be implemented to grasp the key tasks of each river. The pollution sources can be traced based on the river information to banks, then the industry access can be determined by reverse deduction, so as to promote the industrial transformation and upgrading, pipe network construction, sewage collection and other various treatment work. Aiming for improving flood control capacity and water environment and focusing on dredging, sand excavation, river cleaning and control of water and soil loss, the departments shall cooperate with each other and raise fund by multiple channels to carry out the clear water projects steadily.

5. Comprehensively strengthen the water quality safety management of water resources and enhance law enforcement for the protection of drinking water sources. First, the building of qualified rural drinking water source conservation areas shall be fully completed. Second, the management of water works' intake points shall be strengthened, by measures such as establishing a water source inspection system, improving the water production management system and regularly publicizing the quality of drinking water. Third, the water source emergency management and construction shall be reinforced by measures such as preparing emergency plans for water resource allocation, water supply allocation and water resource pollution, establishing teams of professional talents and equipping with relevant emergency equipment and facilities and strengthening emergency drills to form a "triune" emergency disposal and guarantee system integrated with water resource pollution early warning, water source protection emergency disposal and emergency treatment of water works.

6. Earnestly improve a long-term mechanism. All departments shall further strengthen assessment, implement assessment measures; strictly implement the "river segment chief" system with local governments' major leaders as the river chiefs; explore and establish the assessment system for water quality cross-section monitoring in key basins and give full play to the prominent roles of villages and towns in water environment management. A perfect collaboration mechanism for supervision inspection, communication, joint law enforcement and information exchange shall be established to crack down hard on various illegal pollution discharge behaviors. Regular disclosure shall be carried out to strengthen the media and public supervision over the water environment treatment work and a collaboration mechanism for public engagement shall be established and improved. Departments at all levels shall enhance the publicity to create a favorable environment for water control, and relevant departments shall carry out a variety of publicity activities of rich contents. New programs and columns of broadcast, TV or newspapers shall be launched to strengthen the publicity and continuously promote the public awareness of water environment protection, to let the public become water environment protection participants and water environment treatment supervisors, which can fully mobilize the initiative and enthusiasm of vast business owners and give full play to the guiding role of industry associations to form a three-dimensional supervision network and win the "battle" of water control.

3.1.2 Control of combined system overflow pollution

3.1.2.1 Existing major problems

The control of combined sewer overflow (CSO) pollution aims to reduce the total amount of pollutants entering the receiving water by overflow. Due to the complexity of combined sewer system and the randomness and variability during the pollutant transport and overflowing processes, the CSO pollution control involves a series of complicated theoretical and engineering practical problems, so it is necessary to make scientific and systematic decisions to reduce the investment and improve the benefits.

Most of combined systems are in old urban areas and the combined system reconstruction and CSO control measures are faced with many problems such as large construction difficulty, wide affected scope and huge investment, and additionally, there are many restrictions of subjective and objective conditions. As a result, many difficulties and confusions appear during the implementation of CSO pollution control, such as limited to local implementation, single control means, high investment and low efficiency.

The pollutant transport process in the combined sewer system includes many links, such as the collection, transport, interception, storage, treatment, sedimentation in dry season and overflow discharge in rainy season of stormwater and sewage. The operating situations of the system can be roughly divided into two types, namely, sunny days and rainy days (see Figure 3.1).

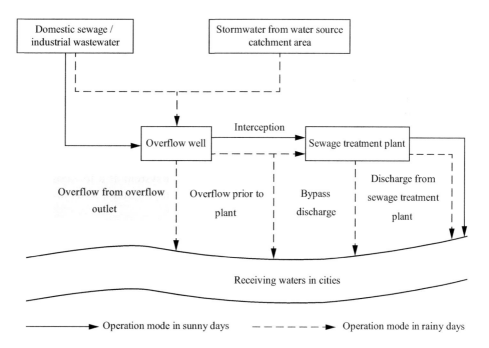

Figure 3.1 **Operation mode diagram of combined sewer system.**

Through the analysis on the entire combined system based on Figure 3.1, the CSO pollution mainly includes three parts: the overflow from overflow wells formed by water exceeding the pipe intercepting capacity, overflow prior to plant formed by water intercepted to sewage treatment plants but exceeding the treatment capacity and the bypass discharge (primarily treated only) of sewage treatment plants. At present, some cities of China mainly pay attention to the CSO pollution control at overflow wells but ignore links such as reasonable interception, overflow control and process matching of sewage treatment plants, and attach no importance to the contribution of source control measures to the CSO pollution control.

The occurrence, transport, load and distribution of the CSO pollution are affected by many factors, such as rainfall characteristics (rainfall amount, intensity, type, etc.), underlying surface conditions (landform, infiltration, retention capacity, etc.), pipe topology and overflow distribution, intercepting capacity (interception ratio), treatment capacity and process configuration of sewage treatment plants.

3.1.2.2 Countermeasures

In order to achieve the goal of efficient reduction of total CSO pollutants, the source control measures shall be adopted, such as by reducing the input to the combined system, appropriately improving the intercepting capacity of intercepting pipes and improving the treatment capacity of sewage plants accordingly, reducing the overflow from overflow wells, or reasonably determining the interception ratio according to the planned and designed capacity of sewage treatment plant to reduce unnecessary investment in main intercepting pipe and overflow prior to sewage treatment plants. As for the combined sewage exceeding the capacity of intercepting pipe and sewage treatment plant, it shall be stored temporarily in regulating storage facilities set up nearby overflow wells and then transported to sewage treatment plants after raining, or utilized or discharged after in situ treatment. As for the combined sewage exceeding the regulation and storage capacities of regulating storage facilities, the overflow of combined sewage shall be discharged after in situ treatment. The sewage treatment plants and their treatment process shall be reasonably upgraded and transformed to reduce the in-plant bypass discharge and improve the sewage treatment efficiency in rainy season.

By the construction of CSO pollution control system, the implementation of source low impact development (LID) or green stormwater infrastructure shall be strengthened to reduce the stormwater entering the combined system at a low cost and also optimize and improve the treatment capacity of the entire system, so as to reduce the COS pollution efficiently to the greatest extent.

1. Source control.
   The combined sewer pollution may be caused by a large amount of stormwater runoff entering combined systems during rain storms. Therefore, the most direct and effective combined sewer pollution control method is to reduce the stormwater runoff entering combined systems from the source. It may be difficult to carry out source control measures in large scale in built-up urban areas to replace all traditional gray infrastructure, but the LID-based CSO pollution control provides a new idea characterized by small investment and high comprehensive benefits, which is worthy of research and popularizing.

2. Reasonable intercepting capacity.

    The intercepting capacity is closely related to the treatment capacity of sewage treatment plants. The interception ratio is an important parameter of the CSO pollution control system. According to the *Code for Design of Outdoor Wastewater Engineering* (GB 50014-2006), the interception ratio should be within the range of 1–5. In engineering practice, it is generally believed that the greater the interception ratio within the said range, the better the CSO pollution control effect. The interception ratio shall be determined reasonably based on the sewage treatment plant capacity and techno-economic analysis.

3. Reasonable design of regulating storage facility.

    In order to further reduce the total overflow pollutant amount, regulating storage facilities are usually needed to be set up in combined systems to temporarily store the combined sewage that exceeds the intercepting capacity of intercepting pipes and treatment capacity of sewage treatment plants during rain storms and, after raining, to transport the sewage to sewage treatment plants for treatment or reuse or discharge into water bodies after in situ treatment. The function principles and scale determination methods for regulating storage facilities in different locations are different. The CSO regulatory control technology is relatively mature with quick implementation and effect, but it requires a relatively large underground space and very high construction, operation and maintenance expenses. Moreover, due to the impacts of factors such as the rainfall characteristics, water collection areas, hydraulic and sediment conditions of pipe systems, setting locations and scale, the reasonable scale of CSO regulating storage pond is the key to the design. Otherwise it will be difficult to ensure the control effect and investment reasonability.

4. Improvement of the capacities of sewage treatment plants.

    It is of great significance to ensure the treatment of intercepted combined sewage for the reduction of total CSO pollutant amount.

    ① Utilization of existing facilities in sewage treatment plants and construction of special treatment facilities for rainy days.

    ② It is a widely applied measure to construct regulating storage ponds in sewage treatment plants for the storage of peak flow.

    ③ CSO in situ treatment. The timely in situ treatment of combined sewage to remove the maximum pollutants such as settleable solids, suspended solids (SS) and bacteria in a short time is a relatively economical, practical and effective method. The layout of CSO in situ treatment facilities is mainly restricted by the urban site conditions, so it is necessary to select appropriate technologies and devices according to the site space conditions. In addition, simple grating interceptor and disinfection facilities near overflow outlets can effectively remove relatively large SS, bacteria and other pollutants, which are characterized by small investment and occupied area and relatively easy implementation.

Besides, the CSO pollution control schemes can be established by following several important principles: ① on the basis of comprehensive survey and full understanding of drainage system conditions, give priority to improving combined systems based on existing facilities and adopting corresponding CSO pollution control measures to reduce CSO pollution and avoid blind and simple adoption of "combined-to-separate" measures; ② widely popularize source green control measures, such as stormwater infiltration, storage, collection and utilization; ③ try to design regulating storage ponds in the pipeline or near overflow outlets or adopt in situ treatment measures and real-time control technology; ④ reasonably intercept the flow and implement corresponding treatment to reduce or avoid the overflow discharge of water transported by main intercepting pipes from a long distance to downstream sewage treatment plants.

3.1.3 Prevention and control of basin water pollution

"Water pollution" refers to the phenomenon that the chemical, physical, biological or radioactive characteristics of water are changed due to the intervention of certain substance, thereby affecting the effective utilization of water, harming the human health or destroying the ecological environment, resulting in the water quality deterioration and posing a serious threat to the survival and development of human beings. Water environment pollution is mainly caused by unreasonable agricultural development or sewage and wastewater discharged into rivers and lakes without treatment or satisfactory treatment. Water pollution in Chongqing is mainly organic pollution, with main pollutants including COD, NH_3-N, *Escherichia coli*, petroleum, etc. Water pollution in the upper reaches of the Yangtze River is mainly caused by domestic waste, industrial wastewater and vessel sewage from cities and towns along the river.

Pollution sources causing water pollution accidents can be divided into physical sources, chemical sources and biological sources according to their attributes, among which chemical sources are the dominant and the most harmful one. Chemicals entering water will cause abnormal chemical properties or chemical concentration in water, resulting in water pollution accidents. According to definitions as specified in the *Identification of Major Hazard Installations for Hazardous Chemicals* (GB 18218-2018), a hazardous substance refers to "a substance or mixture of several substances with a risk of causing fire, explosion or poisoning because of its chemical, physical or toxic properties", and a major hazard refers to "a unit where hazardous substances are produced, processed, handled, used or stored in long term or temporarily and the quantities of hazardous substance are equal to or exceed critical quantities".

1. Domestic waste.

   Domestic waste refers to solid waste generated in daily life or daily activities of offering services, which not only can affect people's daily life but also have a huge impact on water resources necessary for people. Some rural areas do not have suitable waste dumping sites, and the centralized waste dumping will also have a great impact on the surrounding residents' lives, especially on water environment, because many harmful substances in garbage can penetrate soil and directly enter groundwater, causing groundwater pollution. Besides, chemical reactions may occur in some waste piled up for a long time, resulting in harmful substances polluting water sources. Moreover, a part of harmful substances may flow into the nearby rivers and streams along with the rain.

   As for pollution caused by domestic waste, Chongqing has also adopted some strategies. ① The environmental protection publicity is enhanced to raise the people's environmental awareness, and some corresponding rules and regulations are established to intensify the punishment for littering and illegal discharge of factories and individuals and praise those with contributions to environmental protection, so as to promote the implementation of environmental awareness. ② Forces are organized to remove riverside waste and weeds, and teams are set up to dredge rivers to recover clean appearances of rivers. ③ On principles of reduction, harmlessness and reclamation, the management on the whole process of waste generation is reinforced to reduce waste from the source. For waste already generated, harmless disposal and recycling are actively carried out to avoid pollution to environment. ④ The management and domestic waste treatment are intensified, and garbage disposal plants are built in different areas to realize 100% harmless disposal rate of domestic waste.

2. Industrial wastewater.

Chongqing has a high proportion of traditional chemical industries characterized by high energy consumption and heavy pollution. The unreasonable historical industrial layout made it difficult to control industrial pollution in the past. In order to improve the ecological environment and solve water environment problems, Chongqing has adopted effective measures. ① The management on wastewater discharge from industrial parks is reinforced and the reuse efficiency of reclaimed water is increased. ② The green manufacturing is fully implemented to promote the realization of stable up-to-standard discharge of water pollutants. The cleaner production is a scientific method for achieving the maximum resource utilization rate and minimizing the environmental pollution by preventive control of the whole process from production raw materials, production process to product after-sales services. It not only focuses on pollution mitigation and environmental protection but also stresses the water resource saving and economic development. Moreover, by measures such as implementing the whole production process control, continuously adopting improved designs, using clean energy and raw materials, adopting advanced process technology and equipment, improving management and comprehensively utilizing, it can reduce pollution from the source, improve resource utilization rate and reduce the pollutant production and discharge during the production, service and use of products, striving for the most efficient use of resources and minimum discharge of waste (sewage), so as to achieve the harmony, unity and sustainable development of the economy, society and ecology. ③ It has actively implemented water environment integrated treatment and ecological restoration for polluted water bodies.

3. Vessel sewage.

Vessel sewage can be generally divided into domestic sewage and oily sewage: the domestic sewage includes black water and gray water (gray water from kitchen and washing), while the oily sewage includes bilge oily water, oil tank ballast water and tank washing water. As vessel domestic sewage is distinctly different from urban sewage, there are corresponding provisions for the discharge of vessel domestic sewage in China. The maximum allowable pollutant concentrations in domestic sewage discharged into inland rivers in China are as follows: COD no more than 50 mg/L, suspended solids no more than 150 mg/L and coliform group no more than 250 MPN/100 mL. The discharge of vessel domestic sewage exceeding the standards will cause a lot of harm, such as consuming the dissolved oxygen in water, causing red tide, endangering the survival of fish and most of the aquatic life and producing unpleasant odor, which will destroy the beautiful environment, affect benthic organisms, reduce the resistance of aquatic life, reduce the output and introduce cancerogenic substances in the water body food chain and human aquatic food.

As for current vessel sewage situations, specific solutions shall be found and implemented. ① The environmental supervision responsibility shall be intensified. According to departmental rules and regulations for supervision over the Yangtze River vessel sewage, the maritime, fishery and environmental protection departments shall coordinate the supervision to standardize the sewage treatment and disposal. ② Vessels' capacity for sewage prevention and control shall be enhanced. New environment-friendly vessels shall be encouraged, while old and backward vessels shall be replaced or improved. Supporting sewage treatment units of vessels shall be completed and shore installations for sewage receiving and treatment shall be increased. The vessel fuel oil consumption shall be reasonable and the oily sewage generated shall not be discharged until reaching relevant discharge standards. ③ The environmental awareness of enterprises and crew shall be enhanced. Relevant pollution prevention rules and regulations shall be established to let enterprise assume the responsibility for vessel sewage prevention and control. As for the crew, especially key crew members such as chief engineers and boatswains, special environmental protection education and training shall be carried out to make water pollution control start from everyone.

The prevention and control of water pollution are important parts for realizing harmonious development of economy, society and environment protection and are related to the modernization construction of China and vital interests of the broad masses. For prevention and control of the basin water pollution, Chongqing has put forward and timely implemented some pertinent measures, such as reasonably increasing sewage treatment plants, improving urban environmental monitoring and infrastructure information system management, intensifying the industrial wastewater treatment of key pollutant-discharging enterprises in chemical, pharmaceutical, dyeing, food and other industries, carrying out basin ecological compensation pilot projects and continuing to carry out cross-regional cooperation on water environment protection. It has realized a collaboration mechanism for unified planning, standards, evaluation and monitoring for pollution prevention and control regions and basins and has established a water ecological environment function zoning management system. By seeking measures for water pollution control, Chongqing aims to protect the sustainable use of water resources and enable the people to drink clean water.

3.2 Improvement of stormwater and sewage treatment patterns in mountainous cities

3.2.1 Decentralized treatment pattern

At present, the urban areas have been expanding rapidly and related industries and services in the expanded areas have also been developed, forming many decentralized pollution sources, most of which are not included in urban sewage collection systems. The sewage generated or discharged from these decentralized pollution sources is discharged in situ due to lacking of treatment, resulting in increasingly prominent pollution of surface water environment. According to the *Notice of the MOHURD, State Environmental Protection Administration and Ministry of Science and Technology on Issuing Policies on Technologies for Urban Sewage Treatment and Pollution Prevention and Control*, the sewage that is unable to be included into the collection system of urban pipe network shall be treated in situ and then discharged after meeting standards. Therefore, the decentralized sewage treatment pattern has relatively large development potential in the urbanization process of China, which can remedy the sewage treatment problems in expansion areas.

The decentralized sewage treatment pattern means to construct small-scale sewage treatment facilities in relatively small areas, suitable for urban regions with scattered dwellers and remote areas, where centralized sewage treatment is unsuitable due to geographical and economic restrictions and the in situ treatment is selected according to local conditions. In such pattern, the sewage discharged from small-scale areas is collected and treated separately. Meanwhile, the in situ treatment is emphasized and low-cost and sustainable treatment systems are adopted for the sewage treatment and reuse.

3.2.1.1 Analysis on characteristics of decentralized treatment pattern

The decentralized sewage treatment pattern is characterized by flexible construction, realization of in situ treatment to reduce pipe network construction and being conducive to reclaimed water reuse.

1. Flexible construction way and short construction period.
   First of all, compared with centralized treatment, the decentralized sewage treatment is not limited by the scale, which means the treatment unit can be a residential building, an office building, or a factory, a residential community. Due to its small scale, the site selection is less limited by environmental conditions and more choices are available to arrange the facilities on the principle of proximity. Besides, there are also more choices of treatment technologies. Meanwhile, the construction involves relatively simple factors. Therefore, the construction period is relatively shorter, and the investment required is relatively less accordingly.
2. In situ treatment to reduce pipe network construction cost.
   As for centralized sewage treatment, sewage treatment plants and corresponding pipe network are needed for sewage collection and long-distance transport. As pipes are easily blocked due to a large number of precipitable substances in sewage, the design standard requirements for the pipe network are relatively high, indicating high costs. If decentralized sewage treatment is adopted, it does not need a large pipe network system for sewage collection and long-distance transport, thereby greatly reducing pipe network construction and maintenance costs and saving money.
3. Contributing to reclaimed water reuse to improve the utilization rate of water resources.
   At present, more than 400 of 600 cities in China are short of water, of which more than 100 are severely short of water. The water supply of large cities such as Beijing and Tianjin has been in the most severe situation that water demand is close to the exploitable amount of water resources, which means the water shortage will

become more and more serious. Therefore, it is imperative to save water and utilize domestic sewage resources, which also suggests that the decentralized treatment technology of domestic sewage will play a greater role in the utilization of sewage resources. Decentralized sewage treatment contributes to reuse of reclaimed water, mainly reflected in two aspects: first, the treatment process can be used flexibly and reasonably according to water quality of sewage in a small area to make the effluent meet the regional water quality standard; second, decentralized sewage treatment facilities can realize in situ reuse of reclaimed water, avoiding construction of large-scale reclaimed water reuse pipe network.

3.2.1.2 Analysis on technology of decentralized treatment pattern

The technology suitable for urban sewage decentralized treatment pattern mainly includes two categories. The first category is artificial treatment technology that utilizes mechanical and automatic equipment to realize flexible control and regulation of sewage treatment process and accelerate the degradation of pollutants, which can be further divided into the following types.

1. A^2/O process. The A^2/O process refers to the anaerobic–anoxic–oxic biological nitrogen and phosphorus removal process. When sewage enters an anaerobic tank, phosphorus-accumulating bacteria will actively absorb volatile fatty acids in the environment when biodegradable macromolecular organics in sewage are transformed by facultative anaerobes into volatile fatty acids. When sewage enters an anoxic tank, denitrifying bacteria will denitrify biodegradable organics in sewage, thereby achieving the purposes of decarbonization and deoxygenation. When sewage enters an oxic tank, phosphorus-accumulating bacteria will actively absorb dissolved phosphorus in sewage, thereby achieving the purpose of dephosphorization.
2. Oxidation ditch process. In the oxidation ditch process, the DO concentration between two fixed aerators will gradually reduce along the flowing process, and the mixture in an oxic state may gradually transit to the next anoxic state. Therefore, the phenomenon of alternate oxic and anaerobic states will occur in an oxidation ditch, finally achieving the purpose of simultaneous denitrification and dephosphorization.
3. Sequencing batch reactor (SBR) process. The SBR process mainly consists of five phases, namely influent, reaction, sedimentation, drainage, mud discharge and idle period. It integrates functions of aeration tank and secondary sedimentation tank into one tank with functions of water quality and quantity regulation, solid–liquid separation and micro-biological degradation. The SBR has the advantages of simple process, good deoxidization and dephosphorization effects and strong adaptability to inflow quality and quantity changes, but the process has high requirements for automatic control systems and the idle rate of reaction tank and other equipment is high.
4. Biological contact oxidation process. The biological contact oxidation process is a biological sewage treatment technology developed by contact aeration on the basis of biological filters, which consists of four parts, namely oxidation tank, carrier, water distribution device and aeration system. With biofilm purification,

adsorption and retention and food chain effects as its function mechanisms, the biological contact oxidation process has the advantages of both biological filter and activated sludge process.

5. Biological aerated filter. When the biological aerated filter is adopted for sewage treatment, microorganisms will be adsorbed on carrier surfaces, and when sewage flows through the carrier surfaces, pollutants will be oxidized and decomposed under effects of microorganism's absorption of organic nutrients, diffusion of oxygen into the biofilm and bio-oxidation inside biofilm.

6. Membrane bioreactor (MBR) process. The MBR process is a membrane bioreactor organically combining bioengineering and membrane separation engineering. It replaces secondary sedimentation tank in biological treatment with high efficient membrane separation technology, and the effluent quality is equivalent to that after secondary sedimentation tank plus ultrafiltration treatment. Therefore, it nearly can intercept all microorganisms in the bioreactor to minimize the effluent organic pollutant concentration, and it can also effectively remove NH_3-N.

The second category is natural treatment technology. Natural treatment technology is to make use of the soil–plant–water system to purify degradable organics in sewage, recycle nitrogen, phosphorus and other nutrients and water as resources through physical, chemical and biological processes in the system, thereby realizing harmless sewage treatment and utilization of sewage resources. The natural treatment technology can be further divided into the following two types.

1. Land treatment technology.
   It includes four types, namely, slow infiltration, rapid infiltration, underground infiltration and constructed wetland.

   ① Slow infiltration. It is a land treatment technology that sewage is distributed on soil surfaces with plants, so that the sewage is purified in the process of flowing through the soil surface and vertical infiltration within the soil–plant system. In the slow infiltration system, the load of sewage distributed is low and the slow infiltration of sewage through soil results in a long retention period of sewage in surface soil rich in microorganisms, so the water purification effect is good that the effluent can reach reuse standards and no secondary pollution will be generated during the treatment process. However, the slow infiltration technology is restricted by climate conditions. The sewage treatment effect is poor when the temperature is low in winter.

   ② Rapid infiltration. It is to distribute sewage on soil surfaces with good permeability, so that the sewage is purified by a series of physical, chemical and biological effects in the process of infiltration. The rapid infiltration has more stringent requirements for hydrological conditions and soil properties of infiltration sites than other land treatment process types. The rapid infiltration can bear a relatively large hydraulic loading and is less affected by temperature, rainfall and other external conditions. It can operate all year round but will cause a certain amount of secondary pollution during the operation.

   ③ Underground infiltration. It is to distribute sewage into a soil layer 50 cm deep from the ground, with certain structure and good diffusion performance, so that sewage moves toward its surroundings through capillary infiltration under the effect of soil infiltration to meet the treatment and reuse requirements. It has the advantages of no impact on the ground landscape, strong phosphorus and nitrogen removal effects and good effluent quality, and the effluent can meet the discharge and reuse requirements. The sewage distribution system and treatment process of underground infiltration are completed underground, so it will not affect the environment or harm the groundwater due to unpleasant odor or breeding of mosquitoes and flies during the treatment process. The system has strong adaptability that it is less affected by climate factors and is not affected by frozen soil in winter, so it can be implemented in both north and south China.

④   Constructed wetland. The constructed wetland is a sewage treatment technology developed in the latest 20 years, mainly composed of artificial substrate filler and aquatic plants. Its primary purification principle is to utilize the synergistic physical, chemical and biological effects of the substrate–microorganism–plant complex ecosystem to realize efficient purification of sewage through filtration, absorption, sedimentation, ion exchange, plant absorption and microbial decomposition. The sewage treatment effect of constructed wetland is ideal and superior to that of traditional sewage treatment process. In the case of a low influent concentration, the removal rate is generally between 85% and 95% for BOD_5, more than 80% for COD, up to 60% for N and up to over 90% for P. Besides, constructed wetland has a unique environment greening function, and large-scale constructed wetland not only can rapidly increase the green area and eliminate the urban heat island effect but also can provide a beautiful new urban ecological landscape. However, constructed wetland has the disadvantage that it will cause secondary pollution during the operation.

2. Stabilization pond.

The biological stabilization pond is a semi-artificial ecosystem, with its principle of sewage purification similar to the self-purification mechanism of natural waters, in which the biophases include bacteria, algae, protozoa, metazoa, aquatic plants and higher aquatic animals; and abiotic factors mainly include light, temperature, wind, organic loading, dissolved oxygen, pH, nitrogen and phosphorus nutrients. The main ecological characteristic is the symbiotic relationship between bacteria and algae. Under suitable light and temperature conditions, algae use carbon dioxide, inorganic nutrients and water to synthesize algal cells and release oxygen through photosynthesis. Heterotrophic bacteria use the oxygen dissolved in water to degrade organic matters, generate carbon dioxide, NH_3-N and water, which become raw materials for algae to synthesize cells. In the series of reactions, dissolved organic matters in wastewater gradually reduce, algae cells and inert biological residues gradually increase and are discharged along with water. Finally, pollutants are not only removed but also recycled as resources in the forms of aquatic crops and aquatic products, and purified sewage is also reused as reclaimed water resource. The sewage treatment is combined with sewage utilization, realizing the sewage resource utilization in sewage treatment. However, the stabilization pond is greatly affected by temperature during operation, so it may cause unpleasant odor or breeding of mosquitoes and flies, affecting the environment.

3.2.1.3 Construction of decentralized treatment pattern

The construction of a regional sewage treatment pattern can be divided into four steps, namely, the analysis on regional characteristics, analysis on regional sewage discharge characteristics, selection of treatment technology and recommendation of matching technology for different types of regions. Taking the decentralized treatment technical pattern for up-to-standard discharge as an example, this book introduces the construction process in detail as follows:

1. Analysis on regional characteristics.
   The decentralized treatment technical pattern for up-to-standard discharge is primarily applicable to small residences, office buildings, hotels and other regions that are not within the planning scope of urban centralized pipe network. The construction of sewage treatment units in these regions is mainly funded by local parties, so the technology selected must be economical, efficient and easily feasible; the land shortage problem exists in these regions, so the area occupied by the technology selected shall be small; these regions are not far from the sewage treatment unit installation sites, so every effort must be made to reduce the possible impact of odor and noise on users.
2. Analysis on regional sewage discharge characteristics.
   In these regions, the sewage discharge is relatively small and mainly concentrated in the daytime. Most of the sewage is domestic sewage, in which main pollutants are organics and SS, with low content of nitrogen and phosphorus. Therefore, biochemical treatment can be adopted.
3. Selection of treatment technology.
   Main pollutants in such regions are organics and SS, with low content of nitrogen and phosphorus, so enhanced secondary treatment processes such as A^2/O, oxidation ditch, SBR and MBR can be selected; the occupied area of sewage treatment facilities in such regions should not be large, so the SBR and MBR are preferred; the construction of sewage treatment units in these regions is mainly funded by local parties, so the economical SBR is preferred.
4. Recommendation of matching technology for different types of regions.
   According to the above analyses, the biological aerated filter and biological contact oxidation processes can be selected if the Class II discharge standard is implemented, and the SBR process can be selected when the Class I discharge standard is implemented.

3.2.2 Scientific stormwater management[1]

3.2.2.1 Stormwater management measures

It shall be prohibited to stack earthwork construction materials (soil, gravel, concrete, etc.) on or near the ground of permeable pavement. Measures such as the setting of vegetative filter strips, transferring eco-grass ditches and sedimentation tanks shall be taken to reduce runoff or other particles directly flowing from bare soil and guide water with high-impurity content to permeable pavement areas as much as possible, and sedimentation tanks shall be timely cleaned up. As for areas paved with gaps reserved, sediments and debris in gaps shall be timely cleaned up. Fallen leaves in permeable pavement areas shall be cleaned up as soon as possible when they are in a dry state, and the moss on permeable sidewalks shall be timely removed with a hard broom. Vehicles or other equipment exceeding the design load shall not be allowed to enter permeable pavement areas.

The permeability of permeable pavements shall be regularly inspected. As for areas paved with permeable materials, a pavement permeability meter can be used on site for the measurement by variable water head method, and the pavement shall be cleaned up when the penetration rate is below 25 cm/h. As for areas paved with gaps reserved, a certain amount of water can be loaded on a certain area (4–5 m^2) and the time required for complete penetration shall be recorded and compared with the time upon completion of construction to evaluate the permeability.

Soil particles in gaps of permeable pavements can be removed by the following methods: (1) cleaning with high-pressure washing machines (permeable pavement cleaning tanker, etc.); (2) spraying and washing; and (3) blowing with compressed air. If the high-pressure water washing is adopted, the water pressure shall be controlled to the extent not causing damage to the road surface, and the cleaning wastewater with high sediment content shall be properly treated.

The filler shall be dug out for cleaning or replacement if the permeable substrate is blocked. After the permeable asphalt pavement reaches its functional life and needs surface or base course patching, the pavement potholes and cracks can be patched with conventional impermeable asphalt mixture, but the accumulative patched area shall not exceed 10% of the whole permeable area. As for eco-grass bricks and other permeable pavement methods involving plants, the plant disease and pest inspection and prevention and weed removal shall be carried out.

If the permeable pavement area is frequently blocked, possible causes shall be analyzed and reasonable measures shall be taken to eliminate or reduce the impacts. As for damaged permeable pavements, the permeable bricks shall be repaired or replaced with original permeable materials or permeable materials with permeability and other properties not inferior to the original ones. In case of larger-area cracks, potholes, peeling and aggregate peeling of permeable concrete pavements, maintenance must be carried out. Before maintenance, construction schemes for maintenance shall be developed according to damage conditions of the permeable concrete pavements; for the maintenance, the new permeable concrete shall be paved after pavement loose aggregates are removed and dust and debris on road surfaces are cleaned up.

As for permeable pavements with drainage pipes/channels below, regular inspections shall be carried out to check if the pipes/channels are blocked by mud, sand or plant root systems and if they are misaligned or cracked. Repair measures such as jetting cleaning and pipe replacement shall be adopted according to the inspection results. The safety inspection of permeable pavement areas can be carried out by road administrative departments and areas with heavy traffic and dense population shall be taken as key inspection areas.

The reference inspection frequency for each facility structure and item is as shown in Table 3.1.

In order to ensure the operation and maintenance of facilities, if the facilities belong to public projects, relevant administrative departments shall provide working personnel with necessary tools and materials; if the facilities belong to non-public projects, working personnel may prepare tools and materials themselves or apply to relevant departments. Reference materials and equipment are as shown in Table 3.2.

Table 3.1 Inspection and maintenance frequency for permeable pavements

Inspection content	Inspection and maintenance frequency	Item
Permeable pavement area	Earthwork material stacking	N
	Whether there is runoff with high-sediment content entering	2, S
	Sedimentation facilities	S
	Leaves, garbage, debris, etc.	Consistent with municipal sanitation
Ground paved with permeable asphalt, permeable concrete	Permeability inspection	2, S
Ground paved with permeable bricks	Permeability inspection	2, S
	Moss	2
	Damage and missing of permeable bricks	2
Ground paved with permeable bricks, perforated bricks and gravel	Permeability inspection	2
	Inspection on plant diseases and pests, weeds	N
	Damage of perforated bricks, missing or washing away of bricks	2, S
Drainage pipes/ channels below	Blocking, cracking, collapse, crushing, misalignment	2, S
Safety inspection	If there is deformation, damage, crack, collapse, subsidence, pothole, etc. of facilities	12

Notes: 2—every 6 months; 12—monthly; S—after the 24-hour precipitation is equal to or greater than the maximum precipitation every 10 years; F—in season of falling leaves; N—as needed; the inspection and maintenance shall also be carried out if any abnormality is reported by residents.

Table 3.2 List of materials and equipment for maintenance of permeable pavements

Sediment cleaning and permeability recovery	Structure/pipe inspection and maintenance equipment
• Gloves	• Gloves
• Shovel, pry, broom	• Repair tools (trowel, bricklaying trowel, plastering trowel, putty knife, etc.)
• Bucket	• Excavation tools
• Garbage bag, garbage can	• Flashlight
• Pavement permeability meter	• Mirror (for inspecting inaccessible structures)
• Tapeline or ruler	• Pipes for replacement
• Water fender	• Other replacement materials
• High-pressure washer, permeable pavement cleaning tanker	Plant management equipment
• Dynamic rotary sweeper	• Weeding tools
• Compressed air cleaner	• Plant seeds for replanting
	• Irrigation tools

3.2.2.2 *Maintenance of green roofs*

As for newly cultivated plants or under the long-term arid or other severe climatic conditions, attention shall be paid to the irrigation and maintenance of plants in the facilities, with the irrigation interval controlled within 10–15 days, and the irrigation frequency may be properly increased for roofs with thin planting soil or in summer. The plant growth status shall be regularly inspected, and if the vegetation coverage is lower than 90% (unless otherwise required by the design), the following steps shall be taken: (1) identify causes of poor vegetation growth (e.g., if the irrigation or cultivation method is correct) and correct them; (2) check if the cultivated plants are adapted to local climatic conditions and special water content characteristics of the facilities and (3) replace with other plants if necessary, consult urban landscaping administrative departments on the replacement species and follow the green roof plant selection principles.

If excessively dense vegetation in the facilities causes poor stormwater penetration (draining time exceeding 72 hours) or endangers the structure safety, the following steps shall be taken: (1) check if the pruning or other routine maintenance is enough to maintain the appropriate planting density and appearance requirements; (2) check if the cultivated plant types are always excessively dense, and if yes, they shall be replaced with other plants to avoid the problem of continuous maintenance; (3) large plants can be transplanted to areas outside the scope of facilities and (4) small plants can be partly removed.

It is prohibited to use sharp tools during planting soil turning, plant cultivation and other relevant operations to avoid damage to the filtration layer and waterproof layer. The planting soil turning and plant cultivation shall be completed as soon as possible to reduce the soil exposure time. Fallen leaves, garbage and debris in facilities shall be regularly cleaned up, properly disposed or removed. During the soil exposure period, the soil shall be covered with plastic film or other protective layer to avoid soil erosion by rainfall or wind. Regular weeding, pests control and vegetation pruning shall be carried out, and such maintenance for plants of public projects can be carried out by experienced garden workers in accordance with relevant provisions for urban landscaping. In case of obvious erosion and loss of planting soil, the causes shall be analyzed and corrected, and planting soil shall be regularly supplemented to the design thickness. The filtration layer and waterproof layer shall be regularly inspected and shall be instantly repaired if damage or plant root system invasion is found. Sludge, fallen leaves, garbage and debris accumulated in drain outlets or drainage pipe inlets shall be timely cleaned up.

If the waterlogging time of stormwater facilities exceeds 72 hours and the stormwater draining time after the rain exceeds 36 hours, the following steps should be taken to identify and address the causes: (1) check if the drain outlets, drainage pipe inlets and pipes are blocked and clean them up as needed; (2) check if the planting soil is blocked, due to accumulation of overlying deposits, overcompaction, etc.; dig a small hole to observe the soil profile and check the compaction depth or blocking conditions to determine the depth of soil to be replaced or turned over and (3) check if the filtration layer is blocked and timely clean up or replace as needed.

It is prohibited to stack heavy objects in planting areas and other loads (such as pedestrian passing-by) shall be reduced as much as possible. Maintenance personnel shall take relevant measures to distribute the load evenly during maintenance and, unless necessary, shall not enter planting soil areas still in a wet and soft state.

Drainage pipes/ditches shall be regularly inspected and shall be instantly repaired, replaced or corrected in case of damage, crack or misalignment. Retaining walls shall be regularly inspected and shall be reinforced or repaired in case of a crack or breach over 5 cm occurring due to collapse, damage or erosion. Water with the quality not meeting subsequent use requirements shall be reused after treatment.

The reference inspection frequency for each facility structure and item is as shown in Table 3.3.

In order to ensure the operation and maintenance of facilities, if the facilities belong to public projects, relevant administrative departments shall provide working personnel with necessary tools and materials; if the facilities belong to non-public projects, working personnel may prepare tools and materials themselves or apply to relevant departments. Reference materials and equipment are as shown in Table 3.4.

3.2.2.3 Maintenance of bio-retention facilities and sunken greenbelts

When an inlet is unable to effectively collect stormwater runoff on the collection area, the inlet size shall be increased or a local depression shall be set. Garbage and sediments around water inlets and overflow outlets shall be timely cleaned up to ensure unobstructed water flowing. Anti-scouring facilities (e.g., energy-dissipation gravel)

Table 3.3 Inspection and maintenance frequency for green roofs

Item	Inspection content	Inspection and maintenance frequency
Drain outlet	Blocking	2, S, F
	Erosion, damage	2, S
Retaining wall	Crack, settlement, erosion, damage	2, S
Planting soil	Permeability	S, N
	Soil loss	S
Filler	Cleanliness and permeability	2, S
Drainage pipe/ditch	Blocked, damaged, misaligned or not	2
Cleaning of garbage in facilities	If garbage and debris exist in facilities	12
Vegetation	Vegetation survival status	N
	Check the vegetation appearance to determine if pruning is needed	N
	If vegetation has diseases or pests	N
	Weed growth state in facilities	N
	Vegetation is excessively dense	N
Waterlogging	If the waterlogging time exceeds 72 hours	S
Water quality	If relevant water quality requirements are met	2

Notes: 2—every 6 months; S—after the 24-hour precipitation is equal to or greater than the maximum precipitation every 10 years; F—in season of falling leaves; N—as needed; the inspection and maintenance shall also be carried out if any abnormality is reported by residents.

Table 3.4 List of materials for green roof maintenance

Gardening equipment and materials	Erosion control and repair materials
• Gloves • Weeding tools • Soil turning tools (hoe, harrow, shovel, etc.) • Pruning tools (gardening scissors, etc.) • Trolley • Garbage bag, garbage can • Load distributing board • Plant	• Gravel, pebble • Cement • Brick • Repair tools (trowel, bricklaying trowel, plastering trowel, putty knife, etc.)
Irrigation equipment	Temporary covering materials
	• Plastic film • Dust screen
• Hose • Nozzle • Water bag, bucket • Irrigation tank • Water source	Pipe/structure inspection and maintenance equipment
	• Excavation tools • Flashlight • Mirror (for inspecting inaccessible structures) • Tapeline or ruler • Pipes for replacement
Sediment cleaning, planting soil and filler permeability recovery	Special equipment
• Gloves • Shovel • Pressure water gun • Garbage bag, garbage can • Equipment for soil turning and excavation • Planting soil for replacement • Filler for replacement	• Soil monitoring equipment (sampling cutting ring, soil drill, soil nutrient testing kit, etc.) • Water testing equipment

of water inlets and overflow outlets shall be properly maintained to guarantee their design functions.

Garbage and debris in facilities shall be timely cleaned up, and sediments shall be timely cleaned up if the capacity of regulation and storage space is insufficient due to accumulation of sediments. If the capacity of regulation and storage space is insufficient due to the gradient, retaining dams shall be added or the elevation of retaining dam or overflow outlet shall be raised. Slopes and berms shall be reinforced or repaired in case of a crack or breach over 5 cm occurring due to collapse, damage or erosion. The surface planting soil shall be timely supplemented and the covering layer shall be maintained to maintain the thicknesses of surface planting soil and covering layer. As for facilities with distribution pipes/channels and drainage pipes/channels below, the pipes/channels shall be inspected as required, and according to the inspection results, the blockages shall be timely cleaned up to repair closed pipes and channels. It is prohibited to stack heavy objects in the facilities, and other loads (such as pedestrian or light vehicle passing-by) shall be reduced as much as possible. Maintenance personnel shall take relevant measures to distribute the load evenly during maintenance.

If the stormwater draining time after the rain exceeds 36 hours, the following steps should be taken to identify and address the causes: (1) check if the infiltration is obstructed by leaves or debris accumulated at the bottom; if necessary, the leaves or debris shall be cleaned up; (2) check if covered channels (if any) are blocked; if necessary, the covered channels shall be cleaned up; (3) check if the vegetation is so dense that blocks the infiltration; (4) check if other water inflows (e.g., groundwater, illegal connection); (5) check if the facility service area exceeds the design collection area; if yes, necessary flow division control measures shall be taken by consulting the design institution; and (6) if problems are not solved by the above steps 1–4, the obstruction may be caused by the blocking of planting soil due to accumulation of overlying deposits or overcompaction; dig a small hole to observe the soil profile and check the compaction depth or blocking conditions to determine the depth of soil to be replaced or turned over.

Regular weeding, pests control and vegetation pruning shall be carried out, and such maintenance for plants of public projects can be carried out by experienced garden workers in accordance with relevant provisions for urban landscaping. The facility plants shall be timely replanted or replaced, and the vegetation's tolerance to waterlogging and drought shall be considered for the replanting. As for newly cultivated plants or under the long-term arid or other severe climatic conditions, attention shall be paid to the irrigation and maintenance of plants in the facilities.

If the vegetation survival rate is too low (lower than 75%) during the facility operation, the following steps may be taken: (1) identify causes of poor vegetation growth (e.g., if the irrigation or cultivation method is correct) and correct them; (2) check if the cultivated plants are adapted to local climatic conditions and special water content characteristics of the facilities and (3) replace with other plants if necessary and consult relevant data or urban landscaping administrative departments on the replacement species.

If excessively dense vegetation in the facilities causes poor stormwater penetration (draining time exceeding 72 hours) or endangers the structure safety, the following steps shall be taken: (1) check if the pruning or other routine maintenance is enough to maintain the appropriate planting density and appearance requirements; (2) check if the cultivated plant types are always excessively dense, and if yes, they shall be replaced with other plants to avoid the problem of continuous maintenance; (3) large plants can be transplanted to areas outside the scope of facilities and (4) small plants can be partly removed.

The reference inspection frequency for each facility structure and item is as shown in Table 3.5.

In order to ensure the operation and maintenance of facilities, if the facilities belong to public projects, relevant administrative departments shall provide working personnel with necessary tools and materials; if the facilities belong to non-public projects, working personnel may prepare tools and materials themselves or apply to relevant departments. Reference materials and equipment are as shown in Table 3.6.

3.2.2.4 Maintenance of water-retention gardens and stormwater gardens

Entering of domestic sewage and other non-stormwater runoff shall be prohibited; dumping rubbish into the facilities shall be prohibited; garbage and debris in facilities

Table 3.5 Inspection and maintenance frequency for bio-retention facilities and sunken greenbelts

Item	Inspection content	Inspection and maintenance frequency
Water inlet, overflow outlet	Blocking	2, S, F
	Energy-dissipation gravel, etc.	2, S
	Erosion, damage	2, S
Slope, berm, dam	Crack, settlement, erosion, damage	2, S
	Downstream berm leakage	2, S
Planting soil	Permeability	S, N
Filler	Cleanliness and permeability	2, S
Water distribution/ drainage pipe/ channel	Blocked, damaged, misaligned or not	2
Cleaning of garbage in facilities	If garbage and debris exist in facilities	Consistent with municipal sanitation
Vegetation	Vegetation survival status	N
	Check the vegetation appearance to determine if pruning is needed	N
	If vegetation has diseases or pests	N
	Weed growth state in facilities	N
	Vegetation is excessively dense	N
Waterlogging	If the waterlogging time exceeds 72 hours	S
Water quality	If relevant water quality requirements are met	2

Notes: 2—every 6 months; S—after the 24-hour precipitation is equal to or greater than the maximum precipitation every 10 years; F—in season of falling leaves; N—as needed; the inspection and maintenance shall also be carried out if any abnormality is reported by residents.

shall be timely cleaned up; and over grazing, fishing, landfill, tree planting, soil borrowing and crop planting shall be strictly prohibited in the facilities.

Garbage and sediments around water inlets, outlets and overflow outlets shall be timely cleaned up to ensure unobstructed water flowing. Anti-scouring facilities (e.g., energy-dissipation gravel) of water inlets, outlets and overflow outlets shall be properly maintained to guarantee their design functions.

Front ponds/pretreatment tanks with sediments exceeding 50% of the volume shall be timely dredged. If warning signs for preventing misconnection, misuse and drinking by mistake, guardrails and other safety protection facilities or early warning systems are damaged or missing, they shall be timely repaired and improved; pumps, valves and other relevant equipment shall be timely inspected to ensure their normal operation.

Sediments shall be timely cleaned up if the capacity of regulation and storage space is insufficient due to accumulation of sediments, and the dredged sediments shall be properly disposed of. Slopes and berms shall be reinforced or repaired in case of a crack or breach over 5 cm occurring due to collapse, damage or erosion.

As for facilities with drainage pipes/channels below, the pipes/channels shall be inspected as required, and according to the inspection results, the blockages shall be timely cleaned up to repair closed pipes and channels. Regular weeding, pests control

Table 3.6 List of equipment and materials for maintenance of bio-retention facilities and sunken greenbelts

Gardening equipment and materials	Erosion control and repair materials
• Gloves • Weeding tools • Soil turning tools (hoe, harrow, shovel, etc.) • Pruning tools (gardening scissors, etc.) • Trolley • Garbage bag, garbage can • Load distributing board • Plant	• Gravel, pebble • Cement • Brick • Repair tools (trowel, bricklaying trowel, plastering trowel, putty knife, etc.)
	Covering materials for replacement
Irrigation equipment	• Gravel, pebble • Turf
• Hose • Nozzle • Water bag, bucket • Irrigation tank water sources (e.g., sprinkler)	Pipe/structure inspection and maintenance equipment
	• Excavation tools • Flashlight • Mirror (for inspecting inaccessible structures) • Tapeline or ruler • Pipes for replacement
Sediment cleaning, planting soil and filler permeability recovery	Special equipment
• Gloves • Shovel • Pressure water gun • Garbage bag, garbage can • Equipment for soil turning and excavation • Planting soil for replacement • Filler for replacement	• Mini-excavator • Soil monitoring equipment (sampling cutting ring, soil drill, soil nutrient testing kit, etc.) • Infiltration testing equipment • Water testing equipment

and vegetation pruning shall be carried out, and such maintenance for plants of public projects can be carried out by experienced garden workers in accordance with relevant provisions for urban landscaping. Measures for controlling the unpleasant odor and breeding of mosquitoes and flies must be taken in summer. Proper maintenance management for plant recovery growth in summer shall be carried out by removing dead and decaying plants, timely repairing protective slopes and cultivating protective slope plants and transplanting and replanting missing plants to ensure the purification effect of constructed wetlands and beautiful overall appearance. Relevant data or urban landscaping administrative departments can be consulted for the selection of plants. Wetland plants shall be properly harvested according to the plant growth laws, actual growth conditions and design documents. As for newly cultivated plants or under the long-term arid or other severe climatic conditions, attention shall be paid to the irrigation and maintenance of plants in the facilities.

If the vegetation survival rate is too low (lower than 75%) during the facility operation, the following steps may be taken: (1) identify causes of poor vegetation growth (e.g., if the irrigation or cultivation method is correct) and correct them; (2) check if the cultivated plants are adapted to local climatic conditions and special water content characteristics of the facilities and (3) replace with other plants if necessary, and consult relevant data or urban landscaping administrative departments on the replacement species.

If excessively dense vegetation in the facilities affects design functions of the facilities or endangers the structure safety, the following steps shall be taken: (1) check if the pruning or other routine maintenance is enough to maintain the appropriate planting density and appearance requirements; (2) check if the cultivated plant types are always excessively dense, and if yes, they shall be replaced with other plants to avoid the problem of continuous maintenance; (3) large plants can be transplanted to areas outside the scope of facilities and (4) small plants can be partly removed.

If the effluent quality does not meet design requirements, possible causes shall be analyzed and proper measures shall be taken. Water levels in the facilities shall be timely inspected, especially just after planting and in dry season or rainy season, and the water levels shall be adjusted according to the inspection results.

The reference inspection frequency for each facility structure and item is as shown in Table 3.7.

Table 3.7 Inspection and maintenance frequency for water-retention gardens and stormwater gardens

Item	Inspection content	Inspection and maintenance frequency
Water inlet, overflow outlet	Blocking	2, S, F
	Energy-dissipation gravel, etc.	2, S
	Erosion, damage	2, S
	Pump, valve and other relevant equipment	2
Slope, berm, dam	Crack, settlement, erosion, collapse, etc.	2, S
	Berm leakage	2, S
	Warning signs	2
Drainage pipes/ channels below	Blocked, damaged, misaligned or not	2
Cleaning of garbage in facilities	If garbage and debris exist in facilities	Consistent with municipal sanitation
Vegetation	Vegetation survival status	N
	Check the vegetation appearance to determine if pruning is needed	N
	If vegetation has diseases or pests	N
	Weed growth state in facilities	N
	Vegetation is excessively dense	2, N
	If harvesting is needed	Design document, F
Public hygiene	Unpleasant odor	N
	Breeding of mosquitoes and flies	In summer
Water quality	If relevant water quality requirements are met	2
Water level	Water level	S, in dry season

Notes: 2—every 6 months; S—after the 24-hour precipitation is equal to or greater than the maximum precipitation every 10 years; F—in season of falling leaves; N—as needed; the inspection and maintenance shall also be carried out if any abnormality is reported by residents.

Table 3.8 List of equipment and materials for maintenance of water-retention gardens and stormwater gardens

Gardening equipment and materials	Erosion control and repair materials
• Gloves, rain boots • Weeding tools • Pruning tools (gardening scissors, etc.) • Trolley • Garbage bag, garbage can • Hoe, shovel • Plant	• Dam construction materials (soil, brick, concrete, etc.) • Waterproof materials • Repair tools (gloves, trowel, bricklaying trowel, plastering trowel, putty knife, hoe, shovel, etc.) • Energy-dissipation materials (gravel, pebble, etc.)
Sludge, fallen leaves, garbage and debris cleaning up	Pipe/structure inspection and maintenance equipment
• Gloves, rain boots • Net bag • Garbage bag, garbage can • Sewage pump	• Excavation tools • Flashlight • Mirror (for inspecting inaccessible structures) • Tapeline or ruler • Pipes for replacement
Special equipment	
• Water testing equipment	

In order to ensure the operation and maintenance of facilities, if the facilities belong to public projects, relevant administrative departments shall provide working personnel with necessary tools and materials; if the facilities belong to non-public projects, working personnel may prepare tools and materials themselves or apply to relevant departments. Reference materials and equipment are as shown in Table 3.8.

3.2.2.5 Maintenance of stormwater ponds

Garbage and sediments around water inlets, outlets and overflow outlets shall be timely cleaned up to ensure unobstructed water flowing. Anti-scouring facilities (e.g., energy-dissipation gravel) of water inlets and overflow outlets shall be properly maintained to guarantee their design functions. Pumps, valves and other relevant equipment shall be timely inspected to ensure their normal operation. Front ponds/pretreatment tanks with sediments exceeding 30% of the volume shall be timely dredged.

According to actual conditions, the pond appearance and structures shall be inspected monthly, and timely repair shall be carried out in case of any crack, settlement or leakage discovered. Sediments in ponds shall be dredged annually and the dredged sediments shall be properly disposed of. If warning signs for preventing misconnection, misuse and drinking by mistake, guardrails and other safety protection facilities or early warning systems are damaged or missing, they shall be timely repaired and improved. Pipes in or connected to ponds shall be regularly inspected to check if there is any blockage, cracking, breakage, misalignment or misconnection, and maintenance shall be carried out according to the inspection results. The management on sealing and locking of inspection doors shall be reinforced at ordinary times, which means the inspection doors shall not be opened at will and shall be locked immediately after inspection, with records made properly. If the effluent quality does not meet reclaimed water standards, the water shall be properly treated to meet the standards before reuse. The stormwater pond water level shall be monitored in real time while it is raining, and the inlet gate shall be closed to prevent the stormwater entering when the design high level is reached.

The reference inspection frequency for each facility structure and item is as shown in Table 3.9.

In order to ensure the operation and maintenance of facilities, if the facilities belong to public projects, relevant administrative departments shall provide working personnel with necessary tools and materials; if the facilities belong to non-public projects, working personnel may prepare tools and materials themselves or apply to relevant departments. Reference materials and equipment are as shown in Table 3.10.

3.2.2.6 Maintenance of regulating storage ponds

The draining time shall be monitored to check if it reaches design requirements. Garbage and sediments around water inlets and outlets shall be timely cleaned up to ensure unobstructed water flowing. Pumps, valves and other relevant equipment shall be timely inspected to ensure their normal operation. According to actual conditions, sediments shall be timely cleaned up if the capacity of regulation and storage space is insufficient due to accumulation of sediments, and the dredged sediments

Table 3.9 Inspection and maintenance frequency for stormwater ponds

Item	Inspection content	Inspection and maintenance frequency
Water inlet, overflow outlet	Blocking	Monthly, S, F
	Energy-dissipation gravel, etc.	Monthly, S
	Erosion, damage	Monthly, S
	Pump, valve and other relevant equipment	Monthly
Pond wall	Crack, settlement, etc.	Monthly
	Leakage	Monthly
Pipe	Blockage, cracking, breakage, misalignment, misconnection	2
Pond sedimentation	Pond sedimentation situation	1
Water quality	If relevant water quality requirements are met	2
Water level	If the high level is reached	Real-time monitoring while it is raining
Safety inspection	If warning signs are intact	Monthly
	If inspection doors are sealed and locked	Weekly

Notes: 1—annually; 2—every 6 months; S—after the 24-hour precipitation is equal to or greater than the maximum precipitation every 10 years; F—in season of falling leaves; N—as needed; the inspection and maintenance shall also be carried out if any abnormality is reported by residents.

Table 3.10 List of equipment and materials for maintenance of stormwater ponds

Sludge dredging, pond cleaning	Inspection equipment
• Gloves, antiskid rain boots • Garbage bag, garbage can • Sewage pump • Clean water source • Hose • Broom, shovel, etc.	• Flashlight • Mirror (for inspecting inaccessible structures) • Tapeline or ruler
Special equipment	Repair materials
• Water testing equipment	• Pond wall materials (soil, brick, concrete, etc.) • Waterproof materials • Repair tools (gloves, trowel, bricklaying trowel, plastering trowel, putty knife, hoe, shovel, etc.) • Pipes for replacement

shall be properly disposed of. The pond appearance and structures shall be inspected monthly, and timely repair shall be carried out in case of any crack, settlement or leakage discovered.

Water inlet/outlet pipes shall be regularly inspected to check if there is any blockage, cracking, breakage or misalignment, and maintenance shall be carried out according to actual conditions. If warning signs for preventing misconnection, misuse and drinking by mistake, guardrails and other safety protection facilities or early warning systems are damaged or missing, they shall be timely repaired and improved. As for enclosed regulating storage ponds, the management on sealing and locking of inspection doors shall be reinforced at ordinary times, which means the inspection doors shall not be opened at will and shall be locked immediately after inspection, with records made properly.

The reference inspection frequency for each facility structure and item is as shown in Table 3.11.

In order to ensure the operation and maintenance of facilities, if the facilities belong to public projects, relevant administrative departments shall provide working personnel with necessary tools and materials; if the facilities belong to non-public projects, working personnel may prepare tools and materials themselves or apply to relevant departments. Reference materials and equipment are as shown in Table 3.12.

3.2.2.7 Maintenance of eco-grass ditches

When an inlet is unable to effectively collect stormwater runoff on the collection area, the inlet size shall be increased or a local depression shall be set. If soil erosion is caused by water scouring at water inlets, the gravel buffers or anti-scouring measures shall be adopted and properly maintained to guarantee their design functions.

Garbage, sediments and fallen leaves in eco-grass ditches shall be timely cleaned up to ensure unobstructed water flowing, and the sediments and garbage cleaned up shall be properly disposed of. Slopes shall be reinforced or repaired if any collapse, damage or erosion is discovered. Regular weeding, pests control and vegetation

Table 3.11 Inspection and maintenance frequency for regulating storage ponds

Item	Inspection content	Inspection and maintenance frequency
Water inlet, overflow outlet	Blocking	Monthly, S, F
	Energy-dissipation gravel, etc.	Monthly, S
	Erosion, damage	Monthly, S
	Pump, valve and other relevant equipment	Monthly
Pond wall	Crack, settlement, etc.	Monthly
	Leakage	Monthly
Pipe	Blockage, cracking, breakage, misalignment, misconnection	2
Pond sedimentation	Pond sedimentation situation	I
Water quality	If relevant water quality requirements are met	2
Water level	If the high level is reached	Real-time monitoring while it is raining
Safety inspection	If warning signs, guardrails, etc. are intact	Monthly
	If inspection doors are sealed and locked	Weekly

Notes: I—annually; 2—every 6 months; S—after the 24-hour precipitation is equal to or greater than the maximum precipitation every 10 years; F—in season of falling leaves; the inspection and maintenance shall also be carried out if any abnormality is reported by residents.

Table 3.12 List of equipment and materials for maintenance of regulating ponds

Sludge dredging, pond cleaning	*Inspection equipment*
• Gloves, antiskid rain boots	• Flashlight
• Garbage bag, garbage can	• Mirror (for inspecting inaccessible structures)
• Sewage pump	• Tapeline or ruler
• Clean water source	*Repair materials*
• Hose	• Pond wall materials (soil, brick, concrete, etc.)
• Broom, shovel, etc.	• Waterproof materials
	• Repair tools (gloves, trowel, bricklaying trowel, plastering trowel, putty knife, etc.)
	• Pipes for replacement

pruning shall be carried out, and such maintenance for plants of public projects can be carried out by experienced garden workers in accordance with relevant provisions for urban landscaping. The facility plants shall be timely replanted or replaced, and the vegetation's tolerance to waterlogging and drought shall be considered for the replanting. As for newly cultivated plants or under the long-term arid or other severe climatic conditions, attention shall be paid to the irrigation and maintenance of plants in the facilities.

If the vegetation survival rate is too low (lower than 75%) during the facility operation, the following steps may be taken: (1) identify causes of poor vegetation growth (e.g., if the irrigation or cultivation method is correct) and correct them; (2) check if the cultivated plants are adapted to local climatic conditions and special water content characteristics of the facilities and (3) replace with other plants if necessary, and consult urban landscaping administrative departments on the replacement species.

If excessively dense vegetation in the facilities obstructs the water flowing, the following steps may be taken: (1) check if the pruning or other routine maintenance is enough to maintain the appropriate planting density and appearance requirements; (2) check if the cultivated plant types are always excessively dense, and if yes, they shall be replaced with other plants to avoid the problem of continuous maintenance; (3) large plants shall be transplanted to other suitable areas and (4) small plants shall be partly removed.

The reference inspection frequency for each facility structure and item is as shown in Table 3.13.

In order to ensure the operation and maintenance of facilities, if the facilities belong to public projects, relevant administrative departments shall provide working personnel with necessary tools and materials; if the facilities belong to non-public projects, working personnel may prepare tools and materials themselves or apply to relevant departments. Reference materials and equipment are as shown in Table 3.14.

3.2.3 Sewage treatment measures[13]

3.2.3.1 Improving urban drainage pipe networks

For the treatment of urban black and odorous water, the sewer system shall be selected according to local conditions, and the planning and construction of the sewer system shall be coordinated with the overall planning and construction of the urban sewage treatment plants and sponge cities. For pollution point sources that are directly discharged into water bodies, pollutant interception measures shall be taken and the sewage collection system shall be improved to realize complete collection and

Table 3.13 Inspection and maintenance frequency for eco-grass ditches

Item	Inspection content	Inspection and maintenance frequency
Water inlet	Blocking	2, S, F
	Buffer facility	2, S
Slope	Collapse, damage, erosion	2
Space in facilities	Accumulation of sludge, garbage, debris, fallen leaves in facilities	Consistent with municipal sanitation
Vegetation	Vegetation survival status	N
	Check the vegetation appearance to determine if pruning is needed	N
	If vegetation has diseases or pests	N
	Weed growth state in facilities	N
	Vegetation is excessively dense	2, N

Notes: 2—every 6 months; S—after the 24-hour precipitation is equal to or greater than the maximum precipitation every 10 years; F—in season of falling leaves; N—as needed; the inspection and maintenance shall also be carried out if any abnormality is reported by residents.

Table 3.14 List of equipment and materials for maintenance of eco-grass ditches

Gardening equipment and materials	Erosion control and repair materials
• Gloves, rain boots • Weeding tools • Pruning tools (gardening scissors, etc.) • Trolley • Garbage bag, garbage can • Hoe, shovel • Plant	• Dam construction materials (soil, stone, etc.) • Repair tools (gloves, hoe, shovel, etc.) • Trolley • Energy-dissipation materials (gravel, pebble, etc.)

Sludge and garbage cleaning up

• Gloves
• Garbage bag, garbage can
• Trolley
• Broom, shovel, etc.

complete treatment. Besides, the wrong and missing pipe connection shall be checked. As for new urban area with abundant rainfall, the separate sewer system shall be implemented; and the overflow combined sewer system shall be gradually transformed to fully combined sewer system and the overflow ports for direct discharge into rivers shall be gradually blocked. As for cities adopted with separate sewer system, if economic conditions permit, the storage facilities such as storage ponds and stormwater tanks will be gradually established to collect, treat and utilize stormwater runoff, so as to control pollution and reasonably utilize stormwater. In arid and semi-arid areas, if there are relatively complete urban sewage pipe networks and sewage treatment plants, the combined system can be adopted to fully utilize pipes and storage facilities to intercept sewage beyond the treatment capacity of sewage treatment plants, so the mixed sewage after rain can be fully treated in sewage treatment plants, thereby reducing the pollution load in rainy season.

As the black and odorous water mainly occurs in old urban areas, considering the basically fixed underground pipelines, crowded surface structures and narrow roads, the adoption of separate system may cause many realistic questions such as incomplete separation of upstream rain and sewage, huge investment and construction difficulties. Therefore, the selection of sewer system shall not excessively rely on the separate system. Considering the practical factors, most of the old urban areas in many cities will still be in state of coexistence of combined system and separate system for quite a long time.

Most of the old urban areas are adopted with intercepting combined systems, namely overflow control facilities are arranged along river banks and lake shores. The original combined pipes are utilized and intercepting pipes are laid along both banks of rivers to collect the sewage. The specific approach is to set up intercepting wells before main intercepting pipes. The diameter of main intercepting pipe shall be determined according to the dry weather sewage flow and interception ratio, theoretically making the dry weather sewage and initial stormwater enter the main intercepting pipes. When the increased stormwater exceeds the transport capacity of the main intercepting pipes, the excessive water, including part of sewage and stormwater overflow, will flow into rivers. Such drainage system has achieved certain results in the concrete implementation of sewage interception in old urban areas. It makes full use of the original combined pipes and avoids a large amount of complex and difficult treatment work, so it not only reduces the impacts of sewage pipes laid on urban roads on the road traffic and surrounding residents, saves a lot of investment and solves the problem of initial stormwater pollution but also is easy to be implemented, which may be a relatively good sewage collection system for some areas within a certain period of time.

In addition to further standardizing the combined sewer system, the effect of source control and pollutant interception can be further improved during the implementation through the following ways:

1. Considering the inevitable mixed connection of rain and sewage pipes in separate systems, the dry weather sewage pump can be added in the stormwater pump room when there is a stormwater pump room at the end of pipeline, so the water in dry season can be pumped up into the sewage pipe by sewage pump to avoid the stormwater system's discharge into rivers in dry season.
2. Initial stormwater regulating storage ponds can be established. If it is difficult to transform all the original combined systems into separate systems, the interception ratio can be appropriately increased or the initial stormwater regulating storage ponds can be set up, for example, the stormwater regulating storage ponds set up along the Suzhou River in Shanghai, which will have good effects on reducing the overflow and pollutants discharged into rivers.
3. A sponge city system can be established. As the urban non-point source pollution is mainly from pollutants contained in stormwater runoff, the stormwater runoff can be effectively controlled by constructing infiltration facilities and retention facilities, so as to reduce the CSO occurrence. Besides, clearing up the garbage around water bodies is also an important measure to control the non-point source pollution.
4. New sewage collection technologies can be actively adopted. For old urban areas where it is hard to collect sewage by traditional gravity pipes, new technologies can be considered, such as the outdoor negative pressure suction sewage collection system. In the said system, the principle of negative pressure suction is applied and one sewage collection well is set up for every one or more households or areas, the bottom of which is connected with a water-sealed suction pipe to form a water seal in the lower part of the water-sealed suction pipe. Driven by the negative pressure in the negative pressure station, the sewage will enter the collection pipe through the water-sealed suction pipe. If this technology is applied, the shallow-buried pipes can be used to reduce the construction quantities. The practice has proved that it has a good effect on the sewage collection in old urban areas.
5. The monitoring and control of the pipe network operation state can be enhanced, especially real-time monitoring of the water level, flow rate and overflow water quality of main pipe intercepting wells, so as to timely know about the operation state of the intercepting wells and to control the initial stormwater intercepting flow through measures such as the intercepting well valve regulation.

3.2.3.2 *Point source pollution control*

Point source pollution of urban black and odorous water mainly includes urban residents' domestic sewage, industrial sewage, livestock and poultry-scale breeding pollution, etc., and the following measures can be adopted for point source pollution control:

1. Up-to-standard discharge of tail water from sewage treatment plants.
   The centralized treatment pattern shall be adopted for domestic sewage of urban residents to ensure the effluent can stably reach discharge standards and meet control requirements for water environment functional zones and water environment capacity of receiving water bodies. If the effluent quality is superior to Class I(A) standard, the membrane bioreactor (MBR), activated sludge (secondary) + biological aerated filter or constructed wetland advanced treatment process can be adopted.
2. Unification of discharge standards for industrial park sewage and urban sewage.
   As for refractory pollutants, the advanced oxidation process and others can be adopted; as for high-salinity wastewater, combined processes such as the membrane separation (reverse osmosis, forward osmosis) + multiple-effect evaporation can be adopted. Enterprises shall be encouraged to implement cleaner production and reclaimed water reuse, and if necessary, advanced treatment units for advanced oxidation, absorption and membrane technology shall be added to improve the effluent quality.
3. Integrated treatment and utilization of breeding wastewater and livestock and poultry excrement.
   The wastewater discharge of scale livestock and poultry farms shall meet relevant discharge standards. Ecological transformation and utilization of sewage and excrement resources such as separation of excrement and urine, separation of stormwater and sewage, composting and utilization of excrement and sewage in situ treatment shall be advocated in the livestock and poultry breeding. The "anaerobic + facultative" biological treatment process with high efficiency of nitrogen and phosphorus removal can be adopted to reach the discharge standards.
4. Sewage interception and collection system.
   The sewage interception and collection system is widely applicable and has become a foundational project for urban black and odorous river treatment, but the construction must be carried out simultaneously with the construction of roads, river channels, dikes and dams. It is of strong systematicness and works only if the capillary, branch and main pipes are complete, forming a network. As for the decentralized direct discharge from sewage interception pipes built by restaurants, hotels, tourist attractions and agritainment resorts around water bodies in suburban areas, the sewage shall be discharged or reused after treatment by urban sewage treatment plants or self-built sewage treatment facilities. As for the sewage interception and collection systems of new projects, no new sewage outfall shall be added in principle; integrated treatment on existing sewage outfalls shall be carried out, and treatment schemes shall be developed according to the sewage outfall classification on the principles of reuse priority, centralized treatment, relocation and centralization, adjusting way of entering water bodies. Besides, as for urban and rural joint areas, the urban and rural construction shall be planned as a whole,

and the source classification and resource utilization shall be carried for domestic waste. In areas not far from urban sewage pipe networks, the sewage shall be collected into the sewage pipe networks in a centralized manner; in areas far from urban sewage pipe networks, the sewage shall be treated in situ and utilized as resources; in areas of land shortage, buried sewage treatment units can be adopted.

3.2.3.3 Sewage treatment plant upgrading and reconstruction

1. Necessity of sewage treatment plant upgrading and reconstruction.

    ① Retirements of national and local government policies
    With the accelerated development of urbanization and industry, environmental issues, especially urban sewage treatment issues, have become hot research topics in China. However, with a large amount of domestic and industrial sewage flowing into rivers, lakes or groundwater, the increasingly serious water pollution has affected the fishery water and domestic water. Water pollution has become one of important factors restricting the development of China. Therefore, China's discharge standards for sewage treatment plants have become more and more stringent.

    Since the issuance and implementation of the *Discharge Standard of Pollutants for Municipal Wastewater Treatment Plant* (GB18918-2002) in 2002, the number of municipal sewage plants has been increasing. The sewage treatment rate also increased from about 30% in 2002 to about 90% in 2015. The "Ten Measures for Water Pollution Control" and *Discharge Standard of Pollutants for Municipal Wastewater Treatment Plant* (Exposure Draft) released in 2015 put forward time requirements for sewage treatment facilities' implementation of Class I(A) standard and required that all municipal sewage treatment facilities in sensitive areas should meet the Class I(A) standard by the end of 2017. Subsequently, more and more urban sewage treatment plants raised the discharge standards from the Class I(B) standard to Class I(A) standard as specified in the *Discharge Standard of Pollutants for Municipal Wastewater Treatment Plant* (GB18918-2002) or even higher standards in response to the national call for energy conservation and pollution reduction. As stipulated in the *Discharge Standard of Pollutants for Municipal Wastewater Treatment Plant* (Exposure Draft), municipal sewage treatment plants shall implement the Class I(A) standard for effluent discharged into national or provincial key basins, lakes, reservoirs and other closed or semi-closed water bodies. Under the background of new requirements and high standards, the upgrading and reconstruction of sewage treatment plants are inevitable and a peak of upgrading and reconstruction will come.

    ② Serious water environment pollution
    Problems of water environment pollution and water resource shortage have made it more urgent to upgrade and reconstruct sewage treatment plants. From the current water pollution of rivers in China, among monitored cross sections in the national environmental monitoring network for seven major river systems in China, Classes I–III cross sections meeting water quality standards for drinking water source areas only account for 41%, while river

cross sections with water quality inferior to Class V account for 27%. From the current water environment quality of lakes in China, the water quality of 43% of the 28 national key control lakes (reservoirs) is inferior to Class V, and the eutrophication of lakes (reservoirs) is increasingly serious. From the current marine environment quality in China, the offshore marine pollution in China has not been alleviated, and some waters are seriously polluted. From the groundwater quality, the number of cities with a trend of aggravating groundwater pollution is still increasing.

Therefore, the control and raising of sewage discharge indicators can alleviate the pollution of surface rivers and lakes, having great significance to improve the water environment in China.

③   Production and operation restrictions of old sewage plants
The equipment is old and out of repair. During the long-term operation process, the equipment of many sewage treatment plants is damaged to varying degrees due to lacking of proper maintenance and maintenance, seriously affecting the sewage treatment. The aging of equipment caused by various reasons makes the equipment unable to operate normally. Consequently, the effluent of sewage plants can't meet relevant standards for sewage treatment.

The treatment capacity fails to match with treatment requirements. As the situation after urban development was not considered at the design stage of sewage treatment plants in many cities, facilities without enough reserved treatment capacity can't meet current treatment indicators. Especially at the present stage of increasingly higher requirements for water quality, old sewage treatment plants should be upgraded, expanded and reconstructed.

2.   Purposes of sewage plant upgrading and reconstruction.

①   To improve the effluent quality
The effluent quality of sewage treatment plants shall be determined according to environmental function requirements of receiving water, upstream and downstream water purposes, water dilution capacity and self-purification capacity to make the effluent quality meet national or local relevant standards. If the water will be discharged into closed or semi-closed water bodies (including lakes, reservoirs, river estuarine), the total nitrogen (TN) and total phosphorus (TP) concentrations in effluent shall be controlled to avoid the eutrophication. Due to the serious shortage of water resources, many cities are actively promoting the reuse of sewage. If the effluent after secondary treatment is delivered to users as reclaimed water, the effluent quality of sewage treatment plants shall be controlled according to the users' requirements for water quality and national or local relevant standards.

②   To increase the water treatment capacity
With the rapid development of urban economy and urban construction in most cities of China, the water treatment capacity of some sewage treatment plants is insufficient, and the reconstruction based on the original ones is needed to expand the capacity. Such phenomenon often occurs in areas with rapid urban construction.

Taking Chengdu, a rapidly developing city in southwest China, as an example, in Chengdu central urban area, there are nine sewage treatment plants with a total treatment capacity of 1.34 million m³/d. As of June 2014, the actual sewage treatment capacity was 1.5461 million m³/d. Meanwhile, the sewage pipe networks and plants had serious sewage overflow problems, causing serious damage to urban water environment. Seven sewage treatment plants in Chengdu have been operating at full capacity. Therefore, further reconstruction on the basis of original sewage treatment plants' process is needed in Chengdu, so as to meet the requirements for water treatment capacity.

③ To improve the utilization of sludge resources
The pollution and sludge-integrated utilization shall also be considered in the selection of technological process schemes. The sludge output shows an increasing trend with the improvement of sewage treatment facilities. Especially for large sewage treatment plants, the disposal of sludge has become a heavy burden, so the sludge utilization has gradually received attention. On the premise of meeting stabilization and harmless standards, the sludge composting shall be preferred to use the sludge in farmland or green space, or the sludge can be used as building material or energy substance.

3. Principles for sewage plant reconstruction.
As for the reconstruction of built sewage treatment plants, treatment processes of reasonable, mature and reliable technology that can achieve treatment requirements and treatment effects shall be adopted. At the same time, new sewage treatment technologies and processes can be actively and reliably adopted according to treatment plants' local conditions and engineering properties, but pilot plant tests or production tests must be carried out to provide reliable reconstruction parameters before new technologies and processes are adopted in China for the first time. The following three principles shall be followed for upgrading and reconstruction of existing sewage treatment plants:

① Low construction cost, energy conservation, low operating cost and small occupied area.
② Simple operation and management, uncomplicated control links and easy to operate.
③ Adapting to local conditions, based on local characteristics of treatment plants, realizing sewage treatment by stages and levels.

4. Common standards for sewage plant reconstruction.

① Inferior to Class I(B) upgraded to Class I(B).
② Class I(B) upgraded to Class I(A).
③ Class I(A) upgraded to Class IV surface water.

5. Common measures for sewage plant reconstruction.

① Based on the influent quality
According to different influent quality types, sewage treatment plants can be classified into the following two types, namely, sewage treatment plants primarily for treating domestic sewage and primarily for treating industrial wastewater. The former is common in cities, mainly for meeting the people's daily living needs, while the latter one is more common in urban chemical industrial parks, mainly for receiving simply treated industrial effluent.

i. Sewage plants primarily for treating domestic sewage of urban residents.

From the influent quality of sewage plants, the influent quality of such sewage plants is ordinary, without high values in indicators such as COD, NH_3-N, BOD_5, TN, TP and SS. Usually, general conventional sewage treatment processes can meet the effluent quality requirements for such sewage. After sewage plants raise the discharge standards, the effluent after conventional processes previously adopted by the plants may fail to meet the new standards in some indicators in some days. In such case, partial reconstruction can be carried out for the original plants' effluent indicators exceeding discharge standards after new standards are implemented.

If the SS and NH_3-N in effluent indicators exceed the standards, the reconstruction can be carried out based on the original process technological conditions by means of improving screen fineness, setting primary sedimentation tank and appropriately increasing aeration rate. If the COD and BOD_5 in effluent indicators exceed the standards, indicating the organic content in effluent is still higher than the standard discharge values and needs further reduction. The organic removal is often improved by additional biological treatment processes, usually the MBR process, moving bed biofilm reactor (MBBR) process or improved oxidation ditch process is adopted. If necessary, an efficient filter is added to further improve the effluent quality to meet the effluent requirements. If the TN and TP in effluent indicators exceed the standards, the removal of the two items can be improved by increasing carbon sources and dosages of other agents, or the denitrification process can be added based on the sewage plants' original processes and efficient filters can be improved to effectively remove the N and P in water by reasonably controlling operating parameters of filters.

ii. Sewage plants primarily for treating industrial wastewater.

Such sewage treatment plants are more common around urban chemical industrial parks, and various indicators of the received water are greatly affected by the plant effluent. Generally, if some influent indicators of sewage treatment plants exceed the standards, for example, one or two indicators of COD, TN and TP exceed the standards, sewage treatment plants need to have excellent effects of removing the said indicators. In such case, there is no conventional process for the plant reconstruction, which shall depend on the actual conditions.

Taking the reconstruction of a municipal sewage treatment plant in Jiangsu Province as an example, industrial wastewater accounts for more than 80% of the received water. During operation of the original sewage treatment plant, there were problems of poor biodegradability of influent and frequent variation of effluent. In the reconstruction work, mimic enzyme catalysis technology was adopted on the basis of the original treatment processes for advanced treatment. The actual operation monitoring results of effluent showed that the effluent quality could stably reach the Class I(A) discharge standard as specified in GB18918-2002, and the overall COD removal rate of the advanced treatment unit reached at 84.62%.

In view of the above problems, the sewage plant was reconstructed based on its existing conditions. At the end of secondary sedimentation tank, an advanced treatment unit was added in the front part of D-shaped filter to ensure the effluent meeting standards stably; the mimic enzyme catalysis "reactive precipitation" wastewater advanced treatment technology (carboxylation of organic pollutant molecules, then complexation and solid–liquid separation finally) and contact flocculation sedimentation water treatment technology (further sedimentation) were adopted for the advanced treatment. After the reconstruction and a period of debugging operation, the COD removal rate of the sewage treatment plant's biochemical unit reached at 56.07%, and the average removal rate of advanced treatment reached at 84.62%, meeting the reconstruction requirements for Class I(A) effluent, improving the overall COD removal rate and providing a reference for upgrading and reconstruction of sewage plants receiving a high proportion of industrial sewage.

②   Based on main removal effects

i. Mainly for COD removal.

If the COD and BOD_5 in effluent indicators exceed the standards, indicating the organic content in effluent is still higher than the standard discharge values and needs further reduction. The organic removal is often improved by additional biological treatment processes, usually the MBR process, MBBR process or improved oxidation ditch process is adopted. If necessary, an efficient filter is added to further improve the effluent quality to meet the effluent requirements.

The MBR process is adopted as the principal process to replace traditional secondary sedimentation tank and advanced treatment parts with membrane modules, saving more land. In order to ensure the TN removal effect, a post anoxic zone is often set up at the end of bioreactor.

The improved oxidation ditch process is mainly to reconstruct the original oxidation ditch process into a process characterized by A^2/O process. Usually, the existing oxidation ditch is divided into functionally independent anoxic zone and oxic zone, with an anoxic/oxic adjustable section between them, and sometimes in order to improve the TN and TP removal effects, a post denitrification zone is set up at the end of the oxic tank.

ii. Mainly for N and P removal.

If the TN and TP in effluent indicators exceed the standards, the removal of the two items can be improved by increasing carbon sources and dosages of other agents, or the denitrification process can be added based on the sewage plants' original processes and efficient filters can be improved to effectively remove the N and P in water by reasonably controlling operating parameters of filters.

③   Three major measures for upgrading and reconstruction of sewage plants

i. MBR process reconstruction.

Major process flow: coarse screen → pump room → fine screen → grit chamber → membrane screen → MBR → disinfection → effluent.

As for the process, a membrane screen needs to be added for the pre-treatment, and fine screen equipment also needs replacement; in fact, the front part of the MBR is a A^2/O biochemical tank and the rear part is a membrane tank. In addition, a membrane treatment equipment room shall be set up. The sludge concentration in the biochemical tank is improved from the conventional 3–4 to 8–10 g/L, so there are great changes both in the sludge reflux mode and reflux ratio. The removal of TN, NH_3-N, COD_{Cr} and BOD_5 is basically completed in the bio-chemical tank, and at the same time, the SS in effluent approaches to 0. Therefore, the TP in effluent can meet the discharge standard after the biological phosphorus removal and effective chemical phosphorus removal.

At present, the MBR process has been applied in Beijing Xiaohe Sewage Plant (100,000 m^3/d), Beijing Huaifang Sewage Treatment Plant (600,000 m^3/d, underground type), Guangzhou Jingxi Sewage Treatment Plant (100,000 m^3/d, underground type), capacity expansion and reconstruction works of Chengdu No. 3/4/5/8 Sewage Plants (750,000 m^3/d) and Wuxi Shuofang/Meicun/Xincheng Sewage Treatment Plants. Among them, the Chengdu No. 3/4/5/8 Sewage Plants are reconstruction projects, so the limit for TN is still 15 mg/L.

ii. MBBR process reconstruction.
Major process flow: coarse screen → pump room → fine screen → grit chamber → MBBR → secondary sedimentation tank → high-density sedi-mentation tank → filter → disinfection → effluent.

In this process, no reconstruction is carried out for the pretreatment, but a kind of biofilm carrier is added in the existing biochemical tank's oxic zone. The biofilm carrier is characterized by large effective area and suitable for microbial growth, with a specific gravity close to that of water, and a filling rate of 30%–50%. The process has a high volumetric loading and can save land, with a sludge concentration of 6–8 g/L. Meanwhile, it is characterized by good resistance to impact load, stable performance, reliable operation, flexible and convenient operation, so it can be well com-bined with the original system. The subsequent high-density sedimentation tank utilizes the principle of contact flocculation to remove SS, which can ensure the SS stably below 5 mg/L, thereby reducing the TP discharge. As the last guarantee measure, both the high-efficiency filter and sand filter can meet the effluent requirements.

At present, the MBBR process has been applied in the upgrading and re-construction works of Dazhou No. 1 Sewage Plant (80,000 m^3/d), Wuxi Lucun Sewage Treatment Plant (200,000 m^3/d) and upgrading and reconstruction works of Qingdao Licun River Sewage Treatment Plant (170,000 m^3/d).

iii. Reconstruction on the basis of existing processes.
Major process flow: coarse screen → pump room → fine screen → grit chamber → multi-mode A^2/O process → secondary sedimentation tank → high-density sedimentation tank → deep-bed denitrification filter → disin-fection → effluent.

In this process, no reconstruction is carried out for the pretreatment, but the existing oxidation ditch is transformed into a multi-mode A^2/O process. Meanwhile, measures, such as supplementing carbon sources, adjusting division, increasing anoxic zone volume and sludge reflux ratio, are taken to make the TN, NH_3-N, COD_{Cr} and BOD_5 in effluent from the biochemical tank meet standards. The added high-density sedimentation tank is mainly for removing SS and TP, and the deep-bed denitrification filter is mainly for removing TN and also can serve as the last guarantee measure to remove the SS and TP in effluent. Under this condition, the filter medium volume shall meet the needs of the denitrification filter, and the filtration rate should not be too high (usually 6–8 m/h), otherwise it is not favorable for the SS and TP removal. The filtration rate shall be decreased from the original design value to ensure the effluent quality can stably meet the standards.

3.3 Establishment of reclaimed water standards

The water reuse system is a complex non-conventional water supply project, characterized by complex and changeable water source quality, many links and complex composition of treatment process. The water quality safety guarantee has high requirements for research means, technological process and water quality supervision. Therefore, the water reuse system planning, design, operation, management and evaluation shall be improved according to relevant standards and norms to guarantee the utilization safety and improve utilization efficiency.

At present, countries around the world have carried out effective exploration of water reuse, but the standardization work in this field lags behind the engineering practice, and there are still prominent problems such as the lack of important standards, inconsistent terminology, unclear concepts and inadequate coordination among standards. Existing reclaimed water quality standards have been mostly established according to the idea of sewage discharge standards, making the standards difficult to fully describe and evaluate the risks of reclaimed water utilization, causing problems such as incomplete water quality indicators, insufficient basis for the determination of standard values and less consideration of differences in users and utilization conditions. At present, there is still a lack of important standards for the evaluation on reclaimed water utilization efficiency, evaluation on reclaimed water quality and evaluation on reclaimed water treatment technology and process. Therefore, it is urgent to research the standardization system and develop standards suitable for reclaimed water characteristics according to the actual demand.

In addition to water quality standards, the reclaimed water standardization system shall also include standards for reclaimed water terminology, water quality evaluation, risk evaluation, system performance evaluation, management supervision and utilization.

1. New step in the development of China's reclaimed water standards.
   In addition to existing reclaimed water quality standards, relevant institutions are preparing guidelines on management, evaluation and utilization, marking a new step in the development of China's reclaimed water standards. Some national standards being developed currently are as follows:

Water reuse management: water quality management for water reclamation plant

Water reuse evaluation: guideline for reclaimed water treatment technology

Water reuse evaluation: technical guidelines for the treatment and reuse of mine water with high total dissolved solids (TDS)

According to the needs of benefit evaluation of reclaimed water use in China, the Chinese Society for Environmental Sciences' organization standard—*Guideline for Benefit Evaluation of Reclaimed Water Use* (T/CSES 01-2019) was officially published in 2019.

2. Rapid development of international reclaimed water standards.

In recent years, relevant new standards for reclaimed water have been established by international organizations and countries such as the International Standardization Organization (ISO), the World Health Organization (WHO), the European Union and the United States.

In 2013, in order to adapt to the needs of international standardization work and promote standardized development of international business in the water reuse field, the Water Reuse Technical Committee (ISO/TC282 Water Reuse) was approved and established by the ISO in July 2013, with three sub-technical committees under it, namely, the Treated Wastewater Reuse for Irrigation (SC1), Water Reuse in Urban Areas (SC2) and Risk and Performance Evaluation of Water Reuse Systems (SC3). After that, the sub-technical committee for Industrial Water Reuse (SC4) was added in 2017. At present, the ISO/TC282 has 43 active member states and observer states. The ISO/TC282 has published 8 international standards and is developing nearly 30 standards.

In 2015, the United States took the lead to issue the *Framework for Direct Potable Water Reuse* (2015).

In 2016, the European Union published the first reclaimed water utilization guideline—Guidelines on Integrating Water Reuse into Water Planning and Management in the Context of the WFD.

In face of the increasing potable reuse researches, practices and development demand around the world, the WHO released the *Potable Reuse: Guidance for Producing Safe Drinking-Water* for the first time in August 2017, aiming to provide technical guidance for countries all over the world on the portable reuse planning, design, operation, management and system evaluation and gradually guide and standardize the extensive, in-depth and sustainable development of potable reuse of reclaimed water.

3.4 Collaboration among multiple departments and consideration to multiple functions

The collaborative management mode integrating municipal, landscaping, gardening and control departments shall be established to promote each department to better accomplish their own work contents and requirements through cooperation among departments and improve the coordinated utilization of water resources by coordinating functions of landscaping, gardening, flood drainage, stormwater utilization, etc. Among them, the municipal department shall manage the flood drainage and stormwater utilization, while the gardening and landscaping departments shall manage the landscaping and gardening functions, finally forming the mechanism with coordinated cooperation among municipal, landscaping, gardening and control departments and overall consideration to functions of landscaping, gardening, flood drainage, storm-water utilization, etc.

3.5 Whole-process close combination

The concept of sponge city construction shall be reflected in all aspects such as the planning, design, construction, operation and management. In addition, the combination of water treatment measures and long-term clear water measures shall be strengthened.

It can be divided into three parts. The first part is the general requirements to clarify the guiding ideology, define five basic principles, emphasize contents such as systematic treatment, ecological restoration, long-term effects and concerted efforts, and put forward the goals, namely to basically eliminate black and odorous water in built-up areas and realize "long-term clear water" and to eliminate more than 80% of black and odorous water in prefecture cities and to fully realize "long-term clear water" by the end of 2020. The second part is key tasks. Three key tasks shall be determined. The first is to achieve the goal of no black and odorous water by implementing 11 treatment works of four categories, namely, source control to intercept pollutants, endogenous source control, ecological restoration and water circulating to maintain quality; the second is to establish a long-term effect mechanism by tightening river chief system, enhancing law enforcement inspection, strengthening river patrol and inspection and intensifying operation maintenance, to avoid recurring of black and odorous water; and the third is to intensify supervision and inspection to urge local departments to earnestly perform their duties and advance the work. Meanwhile, main duties of each department shall be defined. The third part is guarantee measures. Obvious effects of black and odorous water treatment shall be ensured by strengthening organization and leadership, tightening evaluation and accountability, increasing financial support, reinforcing technical support, optimizing approval process, encouraging public engagement and enhancing credit management.

Meanwhile, the publicity and interaction of water treatment area are also indispensable. Grid management personnel shall be organized to distribute brochures in crowded places such as streets, lanes and parks and popularize the current river treatment situation to the masses to let them know about the water treatment current situation, phased results, instability factors and next plans, so as to strive for the public support and understanding and improve the public satisfaction.

Finally, the precision pollutant interception at the source can be realized through big data analysis. Through online monitoring, the water quality dynamic tracking can be realized and the underground pipe network geographic information system can be established. Nowadays, the water treatment measures of scientific water treatment and intelligent water management have brought good effects for realizing long-term clear water. Water environment treatment is a protracted war and also a technical task, requiring both physical and mental efforts. The so-called combination is to determine working approaches and measures on the basis of following objective laws, full consideration of local conditions and mastering the background to ensure the matching of overall and local situations and combination of symptoms and root cause treatment. In the past, the recurring of black and odorous river was caused by many reasons, such as the lack of overall planning for the upstream and downstream, banks and underwater, ecological deterioration, inadequate water supplement and self-purification capacity, sewage entering rivers and separation problems of sewage and stormwater. Therefore, the pertinence and effectiveness of water treatment can only be improved by sticking to the principle of problem orientation, implementing measures and precision treatment specific to symptoms and establishing normal and long-term control mechanism for integrated treatment of the basin water environment on the basis of fully mastering similarities and differences between rivers and water bodies.

3.6 Establishment of third-party evaluation mechanism

The introduction of third-party assessment in the process of supervision on comprehensively deepening reform will inevitably become a trend and will be widely popularized. The approaches for establishing the third-party evaluation mechanism of comprehensively deepening reform mainly include: first, defining the legal status of the third-party evaluation and establishing the authority of the third-party evaluation to legally ensure the legitimacy of third-party evaluation agencies and grant the evaluation right to third-party evaluation agencies, so as to guarantee they can carry out the evaluation independently, objectively and scientifically. The indispensability of third-party evaluation in comprehensively deepening reform shall be guaranteed by making the evaluation a basic link and essential component of the reform evaluation to form a long-term third-party evaluation mechanism. Second, implementing the exposure of decisions. As for reform plans, reform measures or reform projects that involve vital interests of the people and need to be widely known by the public, the decision basis and process shall be published to the public as much as possible before the decision-making to solicit opinions of the public, experts and scholars. Rules and regulations specifying relevant departments' responsibilities and obligations of cooperation on the third-party evaluation shall be issued to promote the implementation of reform policies. Third, establishing an industry supervision system for the third-party evaluation to standardize the third-party evaluation work. The qualification management for third-party evaluation agencies shall be intensified by establishing eligibility criteria for third-party evaluation agencies, and the application examination and approval and dynamic evaluation shall be carried out for relevant agencies. The third-party evaluation process shall be standardized by specifying the evaluation scope, contents, forms, methods and steps to ensure the objectivity and accuracy of evaluation results. Fourth, improving the utilization system for the third-party evaluation results and establishing scientific and reasonable reward and punishment systems and measures to promote the orderly and sustained reform.

Chapter 4

Typical measures for integrated treatment

4.1 Source treatment measures[1]

The construction of sponge city and treatment of black and odorous water may be carried out at the same time and promote each other as they share so many commonalities in the concept and construction way. The stormwater runoff control is an important part in the construction of sponge city, usually realized by means of natural water bodies, multi-functional regulation and storage water bodies and regulating storage ponds constructed by artificial measures. Water bodies with the black and odorous problems solved can be used as natural stormwater regulating storage ponds to solve the land problem for special stormwater management and storage facilities in the sponge city construction.

In the urban black and odorous water treatment, it is not only required to fully eliminate black and odorous water but also needs to scientifically explore supplement water sources by multiple channels, improve hydrodynamic conditions, restore water ecosystem, improve natural purification capacity of water bodies, realize continuous improvement of urban water environment and maintain long-term effects.[4][5]

As for various low-impact development technologies adopted during the construction of sponge city, the requirements for non-point source pollution control of initial rain and the "source control and pollutant interception" in the black and odorous water treatment are completely the same. The annual total settleable solid (SS) removal rate of low-impact development stormwater system can reach 40%–60%, thereby reducing the pollution load of water bodies. In order to achieve long-term effects after the black and odorous water treatment, necessary measures shall be adopted, such as sponge city construction, intensifying the retention, storage and purification of urban rainfall runoff and improving the water liquidity and environmental capacity. Measures such as construction of constructed wetlands and ecological banks adopted in the black and odorous water treatment are also one of the methods for ecological restoration and protection in the construction of sponge city.

By adopting the construction concept of sponge city and combining with the black and odorous water treatment measures, Li Junfei et al. constructed the stormwater collection system, purification system, wetland treatment and water ecological restoration system, stormwater management and storage and infiltration system and built the Yayong River into an urban landscape river with multiple functions, including the restoration of ecological water body, regulation, storage and infiltration of stormwater and connection of Guangdong–Hong Kong green corridor, achieving significant environmental benefits. In summary, the sponge city construction and black and odorous water treatment have common construction demands in aspects such as the runoff pollution control, regulation, storage and utilization of stormwater and water ecological protection. Therefore, during the construction of specific projects, the organic combination of the two can save the project costs and also maximize the project benefits.

4.1.1 Permeable pavement

The permeable pavement is an important source control technology under the concept of "sponge city". Generally, the permeable surface is formed by permeable paving materials such as permeable bricks, permeable asphalt, cobblestones, grass planting bricks and gravel, or formed by traditional materials paved with gaps reserved. As for the pavement with gaps reserved in a narrow sense, the hollow area shall be greater than or equal to 40% of the total pavement area. Characterized by wide applicable scope and convenient construction, this technical measure can replenish groundwater and also can reduce peak flow and purify stormwater to a certain extent. At present, the permeable pavement system has been widely applied in areas such as parks, parking lots, sidewalks, squares and light-load road. Its main functions are to collect, store and treat stormwater runoff, so as to replenish aquifer by infiltration, which is of great significance to improve the overall hydrological regulation and storage function of a city.

The structure of permeable pavement shall meet the provisions as specified in *Technical Specification for Pavement of Water Permeable Brick* (CJJ/T188), *Technical Specification for Permeable Asphalt Pavement* (CJJ/T190) and *Technical Specification for Pervious Cement Concrete Pavement* (CJJ/T135), while the permeability coefficient, appearance quality, dimensional deviation, mechanical properties and physical properties of permeable bricks shall meet the provisions as specified in the current industry standard—*Technical Specification for Pavement of Water Permeable Brick* (CJJ/T188-2012). The strength grade of permeable brick shall be determined by design and the surface course shall be in harmony with the surroundings. The selection of brick type and pavement form shall be determined by designers according to the pavement location and functional requirements. The permeable brick material and its structure shall meet the requirements for high permeability, excellent water retaining capacity, low evaporation, convenient cleaning and maintenance and re-usability. The soil foundation shall be stable, compact and homogeneous with enough strength, stability, non-deformability and durability. The compaction degree of soil foundation shall not be lower than the requirements as specified in the *Code for Design of Urban Road Subgrades*.

The section and real picture of typical permeable pavement are as shown in Figures 4.1 and 4.2, respectively.

If the permeable pavement technical measure is used for the construction of mountainous sponge city, necessary measures shall be taken to prevent the occurrence of secondary disasters or groundwater pollution, especially in steep areas prone to collapse and landslides, areas with special soil geology such as collapsible loess, expansive soil and high-salt soil, and areas with serious runoff pollution such as gas stations and wharfs.

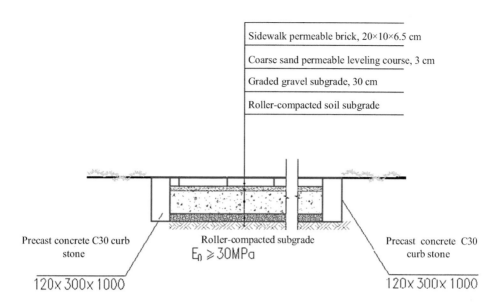

| Sidewalk permeable brick, 20×10×6.5 cm |
| Coarse sand permeable leveling course, 3 cm |
| Graded gravel subgrade, 30 cm |
| Roller-compacted soil subgrade |

Precast concrete C30 curb stone

Roller-compacted subgrade
$E_0 \geqslant 30MPa$

Precast concrete C30 curb stone

120×300×1000

120×300×1000

Figure 4.1 Section of typical permeable pavement.

Figure 4.2 Real picture of typical permeable pavement.

4.1.2 Stormwater management and utilization measures

Due to the rapid urbanization development, the original natural ground capable of water conservation is replaced by large areas of impermeable surface, which significantly changes the natural drainage way and drainage layout and also changes the hydrological mechanism in natural state, primarily reflecting in increased runoff generated by extreme rainfall events, shorter concentration time and greater burden of urban drainage system. Besides, the urban stormwater runoff contains many pollutants such as SS, grease, organic carbon, nutrients, heavy metals, toxic organics and pathogens that are generated by human activities or during natural process. The pollution process of stormwater runoff is relatively complex and mainly affected by factors such as the rainfall and land use surface characteristics. Therefore, the low impact development (LID) technologies applied in sponge city construction are of great significance for the control of total runoff volume, reduction of runoff pollution and utilization of stormwater resources. At present, there are three common measures applied in the stormwater management and utilization, namely, stormwater garden, stormwater pond and regulating storage pond.

The stormwater garden refers to the engineering facility that can infiltrate and filter stormwater by using soil and plants to purify and retain stormwater, so as to reduce the runoff, which is a facility for ecologically sustainable storm flood control and stormwater utilization. The stormwater gardens in mountainous sponge city communities have many functions such as stormwater management, water purification, stormwater resource utilization and restoration of water cycle.

In order to reduce the runoff pollution in communities, stormwater gardens can be built in low-lying parts of community lawns, which shall be kept away from utilities pipelines (especially gas and gravity flow pipelines). Stormwater gardens shall be connected with stormwater drainage ditches around buildings to collect roof stormwater and water drained from green spaces and roads in surrounding areas of the gardens, aiming for the retention, slow drainage, evaporation and plant purification of stormwater, contributing to improving the removal rate of pollution load and control rate of total runoff volume. The section of typical stormwater garden is as shown in Figure 4.3, and the real pictures are as shown in Figures 4.4 and 4.5.

The stormwater pond is an infiltration depression or pond that filters stormwater by natural or artificial pond or depression to supplement groundwater, which can effectively reduce the peak runoff. Combined with green space, open space and other site conditions, the stormwater pond can be designed as a multi-functional water body for regulation and storage, namely to perform the normal landscape, leisure and recreation functions at ordinary times and perform the regulation and storage functions in case of rainstorms to realize the multi-functional utilization of land resources.

Usually, a stormwater pond consists of water inlets, front pond, main pond, overflow outlets, protective slope, revetment, maintenance channel, etc. Stormwater ponds shall meet the following requirements:

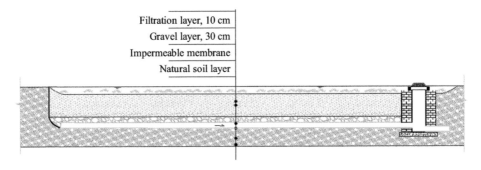

Filtration layer, 10 cm
Gravel layer, 30 cm
Impermeable membrane
Natural soil layer

Figure 4.3 Section of stormwater garden.

Figure 4.4 Real picture 1 of stormwater garden.

Figure 4.5 Real picture 2 of stormwater garden.

1. Energy dissipation facilities such as gravel and energy dissipation berm shall be set at water inlets and overflow outlets to avoid water erosion.
2. The front pond is a pretreatment facility to settle large-particle pollutants in run-off, the bottom of which is usually made of concrete or block-stone structure for the convenience of dredging. The front pond shall be set with dredging channel and protective facilities. The revetment should be ecological soft revetment and the side slope gradient (vertical: horizontal) is usually 1:2–1:8. The sediment area

volume of front pond shall be determined based on the dredging cycle and SS pollution load in stormwater runoff.

3. The main pond usually consists of the permanent volume and storage volume below the normal water level. The water depth of permanent volume is usually 0.8–2.5 m, while the storage volume is usually determined based on the "control volume per unit area" proposed in relevant local planning. The stormwater pond with the peak flow reduction function also includes an adjustable volume, which shall be drained within 24–48 hours. An aquatic plant area (stormwater wetland) should be set between the main pond and front pond. The main pond revetment should be ecological soft revetment and the side slope gradient (vertical: horizontal) should not be greater than 1:6.

4. The overflow outlets include overflow riser and spillway. The drainage capacity shall be determined based on the drainage capacity of the downstream storm sewer or excessive storm runoff discharge system.

5. Safety protection and warning measures such as guardrails and warning signs shall be set up for the stormwater pond.

Applicability: the stormwater pond is applicable to buildings, communities, urban green spaces, squares and other sites with space conditions.

Advantages and disadvantages: the stormwater pond can effectively reduce the total runoff volume, runoff pollution and peak flow and is an important part of the urban waterlogging prevention and control system. However, it has strict requirements for site conditions and costs high for the construction and maintenance. The typical stormwater pond is as shown in Figure 4.6.

As a means of flood retention and stormwater pollution control, the regulation and storage of stormwater are widely used worldwide. Initially, regulating storage ponds

Figure 4.6 **Stormwater pond.**

were only used to temporarily store excessive stormwater, and natural ponds or depressions were always used to store water. With the people's increasingly deep understanding of storm floods and non-point source pollution, the functions and types of regulating storage ponds are gradually diversified. According to the engineering functions, regulating storage ponds are mainly classified into three types: peak discharge regulation, non-point source pollution control and stormwater utilization. In the construction of communities in mountainous sponge city, these ponds can effectively control the annual runoff discharge and realize utilization of stormwater resources. The regulating storage pond can be set in the middle or at the end of a drainage system, and are usually set prior to the regulating storage pond. The section of regulating storage pond is as shown in Figure 4.7. Besides, drop manholes may also be adopted in the construction of stormwater regulating storage pond to slow down the stormwater, so that part of pollutants in stormwater can be settled during the process. The four-level waterfall landscape as shown in Figure 4.7 is an example.

4.1.3 Bio-retention measures[6]

Bio-retention measures are typical LID technologies with the superiorities of reducing urban stormwater runoff, purifying stormwater quality and replenishing groundwater, and also play an important role in mitigating the urban heat island effect, lowering atmospheric temperature, increasing humidity, enhancing biodiversity, etc. Sunken greenbelts, bio-retention facilities and eco-grass ditches are common measures in the construction of sponge city.

The sunken greenbelt refers to a green space with the level no more than 200 mm below the surrounding paved surface or roads, which has a certain regulation and storage volume and can be used to regulate, store and purify stormwater runoff. Sunken greenbelts can collect stormwater runoff generated by surrounding hardened surfaces and intercept and purify small-flow stormwater runoff by the combined effects of vegetation, soil and microorganisms, while the stormwater exceeding the storage and infiltration capacity will be discharged into the stormwater pipe network via gutter inlets. On the one hand, sunken greenbelts can reduce the runoff volume and mitigate

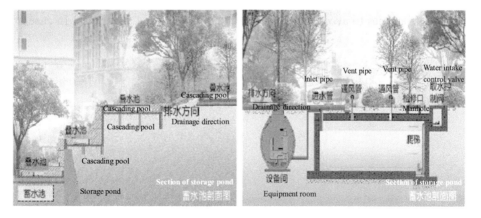

Figure 4.7 Section of regulating storage pond.

urban flood disasters. On the other hand, the infiltrated stormwater can increase the soil moisture content, thereby reducing the irrigation water consumption of green space and also contributes to the conservation of groundwater. Sunken greenbelts are widely used in urban buildings, communities, roads and squares.

The sinking depth of a sunken greenbelt can be determined according to the water resistance of plants and soil permeability, usually 100–200 mm. Meanwhile, overflow outlets (such as gutter inlets) shall be set to ensure the overflow discharge of runoff in case of rainstorms, the top elevation of which is usually 50–100 mm above the greenbelt. The section of typical sunken greenbelt is as shown in Figure 4.8.

The sunken greenbelt is characterized by wide applicable scope and low construction and maintenance costs, but its large-area application is easily affected by terrain and other conditions, and the actual regulation and storage volume is relatively small. In areas with serious runoff pollution, if the bottom infiltration layer of a sunken greenbelt facility is less than 1 m away from the seasonal maximum groundwater level or lithosphere, and less than 3 m (horizontal distance) away from building foundations, necessary measures shall be taken to avoid secondary disasters. Real pictures of typical sunken greenbelts are as shown in Figures 4.9 and 4.10.

Bio-retention facilities refer to the facilities in low-lying areas that can store, infiltrate and purify stormwater runoff by the system of plants, soil and microorganisms, mainly collecting stormwater runoff from adjacent roadways and sidewalks, with the section from top to bottom consisting of water-retention zone/gravel barrier strip, planting soil layer, sand filter layer and pebble layer. Bio-retention facilities are applicable to surrounding green spaces of roads and parking lots in buildings and communities, as well as urban green spaces such as urban road greenbelts. In areas with serious runoff pollution, if the facility's bottom infiltration layer is less than 1 m away from the seasonal maximum groundwater level or lithosphere, and less than 3 m (horizontal distance) away from building foundations, a complex bio-retention facility with anti-seepage bottom may be adopted. The section of typical bio-retention facility is as shown in Figure 4.11.

Due to the large gradient in mountainous city, the road longitudinal gradient shall be considered when bio-retention facilities are applied to road greenbelts. Besides, retaining dams/berms shall be set to slow the stormwater down and increase the stormwater infiltration. Anti-seepage treatment shall be carried out for the facilities' parts close to subgrades to avoid impacts on the stability of road subgrades. In case of the

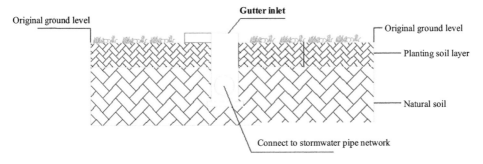

Figure 4.8 Section of typical sunken greenbelt.

Planting soil
Impermeable membrane
Sand cushion
Natural soil layer

Figure 4.9 **Real picture 1 of typical sunken greenbelt.**

Figure 4.10 **Real picture 2 of typical sunken greenbelt.**

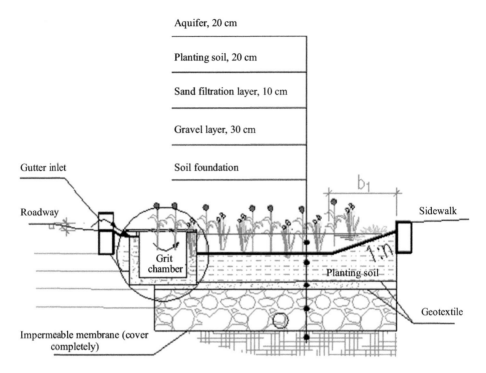

Figure 4.11 Section of typical bio-retention facility.

minimum longitudinal gradient ≤2%, the bio-retention belt may be without retaining dam, but with local bumps of planting soil every 10 m to form terraced micro water storage units; in case of the road longitudinal gradient of 2%–7%, terraced stormwater bio-retention belt shall be adopted; in case of the road longitudinal gradient ≥7%, terraced drop bio-retention belt shall be adopted with retaining dams arranged every 5 m. In order to enhance the water retention, a small retaining dam shall be set between two retaining dams, with the dam top at the same level of the sand filter layer.

The flowchart of a typical road bio-retention belt system is as shown in Figure 4.12, specifically as follows: the road stormwater flows into grit chambers via side wall gutter inlets of curbs and overflows through stormwater gratings of grit chambers, then flows through pebble areas to realize uniform water distribution and re-filtration and finally flows into planting areas where the stormwater is purified under the combined effects of plant, soil and microbial system. After that, the purified stormwater will be collected by blind pipes and drained into existing municipal stormwater system. If the stormwater exceeds the capacity of bio-retention belts, the excessive stormwater will be directly drained into existing municipal stormwater system via stormwater overflow outlets.

Bio-retention facilities are characterized by diverse types, wide applicable scope, convenient combination with landscape, good runoff control effects and low construction and maintenance costs. However, in areas with high groundwater level and lithosphere, poor soil permeability and steep terrain, necessary measures such as soil

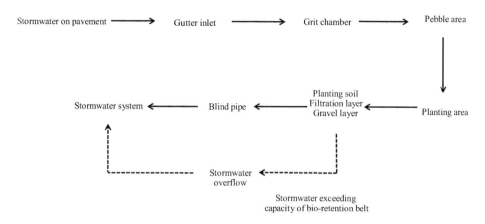

Stormwater on pavement ⟶ Gutter inlet ⟶ Grit chamber ⟶ Pebble area

Stormwater system ⟵ Blind pipe ⟵ Planting soil / Filtration layer / Gravel layer ⟵ Planting area

Stormwater overflow ⟵ Stormwater exceeding capacity of bio-retention belt

Figure 4.12 **Flowchart of typical road bio-retention belt system.**

Figure 4.13 **Real picture of typical bio-retention belt.**

replacement, anti-seepage treatment and terraced construction shall be taken to avoid secondary disasters, resulting in increased construction cost. The real picture of typical bio-retention belt is as shown in Figure 4.13.

Eco-grass ditches refer to vegetated surface ditches that can collect transport and discharge stormwater runoff. Through collecting stormwater runoff by gravity flow, eco-grass ditches are capable of reducing and purifying runoff of impermeable underlying surfaces and also can connect other various facilities, urban storm sewer system and excessive storm runoff discharge system. In addition to eco-grass ditches for transporting stormwater, there are also permeable dry eco-grass ditches and wet eco-grass ditches always with water, which can improve the total runoff volume and the runoff pollution control effect, respectively. This technical measure is applicable to the surrounding areas of roads, squares, parking lots and other impermeable surfaces in buildings and communities and areas such as urban roads and urban green spaces and also can be used as the pretreatment facility of LID facilities such as bio-retention facilities and wet ponds. Eco-grass ditches can be applied in combination with storm sewers and can also replace storm sewers where the site vertical planning allows and the safety is not affected.

Eco-grass ditches can be easily combined with the landscape. The cross section of a shallow ditch should be inverted parabolic, triangular or trapezoidal. The side slope gradient (vertical: horizontal) should not be greater than 1:3 and the longitudinal gradient should not be greater than 4%. In case of a relatively large longitudinal gradient, terraced eco-grass ditches or energy dissipation berms in the middle should be arranged. The maximum flow velocity shall be less than 0.8 m/s and the Manning coefficient should be 0.2–0.3. The height of vegetation in eco-grass ditches for transporting stormwater should be within 100–200 mm.

The real picture of typical eco-grass ditch is as shown in Figure 4.14.

4.2 Water treatment measures

4.2.1 Endogenous pollution treatment measures

The ecological restoration of lakes mainly includes two strategic steps, namely, control of pollution sources and ecological restoration, while the control of pollution sources mainly refers to the control of endogenous pollution and exogenous pollution of lakes. The endogenous pollution mainly refers to the phenomenon that nutrients entering lakes gradually deposit on the sediment surfaces of lakes under various physical, chemical and biological effects and then are released into the water after accumulating to a certain amount. Different from exogenous pollution of lakes, the endogenous

Figure 4.14 Real picture of eco-grass ditch.

pollution of lakes manifests itself after a period of time, because the pollutants in lake sediments float upward with sediments after the exogenous pollution is controlled. Under static conditions, lake sediments can release about 10,000 tons of nitrogen and 900 tons of phosphorus every year, equivalent to 30%–40% of exogenous pollution load. The nitrogen, phosphorus and other nutrients accumulated on sediment surfaces, on the one hand, can enter the food chain and participate in the circulation of aquatic ecosystems through direct ingestion by microorganisms and, on the other hand, can be released from sediments into water again under certain physical, chemical and environmental conditions, thereby forming pollution loads in lakes. There is a collection-to-source transformation process for sediments' receiving of exogenous nitrogen and phosphorus, namely the nitrogen and phosphorus in sediments begin to be released into water with the continuous accumulation of exogenous pollutants. In such case, the endogenous pollution will prevent the water quality from improvement in quite a long time even if the exogenous pollution is cut off.

The endogenous pollution treatment includes physical and chemical treatment remediation and sediment in situ bio-remediation technologies. Commonly used technologies of physical remediation include ecological dredging and sediment in situ capping. The ecological dredging technology means to remove pollutants in water sediments by sediment dredging for the purpose of treatment of endogenous pollution in lakes and rivers, so as to reduce the release of sediment pollutions into water and create conditions for the restoration of aquatic ecosystems. In ecological dredging, the sediment pollution degree will be analyzed and then the ecological, economic and social benefits will be taken into overall consideration, which is different from maintenance dredging. The ecological dredging can effectively remove the endogenous pollution but will also cause certain destruction of benthic habitats, so the rebuilding of benthic habitats and planting of aquatic plants later will be needed to restore the benthic ecosystems. The sediment in situ capping technology refers to the physical remediation method that isolates polluted sediments from water by capping to avoid the transport of sediment pollutants to water, which has a good effect on the aspect of polluted sediment remediation and can effectively prevent harmful substances in sediments from entering water and forming secondary pollution. The key to the sediment in situ capping technology is the selection of capping materials. At present, the commonly used materials include unpolluted sediment, sand, gravel, artificial synthetic material, etc. The sediment in situ capping technology can effectively inhibit the release of sediment pollutants and improve the water transparency, but it does not solve the sediment pollution problem fundamentally and is difficult to be widely popularized due to its high cost.

The principle of chemical remediation is to add chemical agents for the hydrolysis and hydration reactions with pollutants in sediments to turn the free water in pores into bound water solidifying soil, so as to agglomerate small-size solid particles in sediments into larger floccules to separate solid phase from liquid phase. For another, the refractory organics and metallic elements will be converted to stable states by chemical reactions, and the organics, heavy metal ions and other harmful substances in sediments will be effectively stabilized, so as to prevent the organics and heavy metals from being transported, converted and released. After the solidification and stabilization of sediments, the pollutants in sediments will be converted into an immovable state, which can reduce the harm to environment. The solidified and stabilized sediments may be stored and disposed in situ or cleaned up according to demands of the project planning.

The sediment in situ bio-remediation technology refers to the in situ bio-remediation of polluted environment objects without transfer or transport under the conditions of basically not destroying the natural environment of water sediments. The microbial reproduction in sediments can be promoted by adding nutrients required for microbial growth or increasing microorganisms with special affinity. Under the effect of microorganisms, organics will decompose rapidly to thicken the aerobic layer of sediments and accelerate the release of micro-nutrients in sediments, which can block the release of toxic substances from the underlying black and odorous sediments and improve the pollutant decomposition ability of aerobes on the mud–water interface, so as to improve the self-purification capacity of water. The sediment bio-remediation technology can improve the self-restoration capacity of water and maintain the natural balance of water, but it requires a long treatment period and is only applicable to sediments with high organic content.

To sum up, there are three major technologies for sediment pollution control, namely, sediment dredging, in situ covering and in situ passivation. Among them, the in situ covering is also called blocking, masking or capping, and its technical core is to cover polluted sediments with some materials of good blocking effects, so as to separate sediment pollutants from the overlying water, which greatly reduces the ability of sediment pollutants to be released into water. The core of in situ passivation is to add artificial or natural substances with passivation effect to pollutants to make the sediment pollutants in an inert and relatively stable state, thereby reducing the release of sediment pollutants into water and effectively controlling the endogenous pollution. The three technologies have their own advantages and disadvantages, as briefly summarized in Table 4.1.

Research shows that after the exogenous pollution is cut off and treated, pollutants in sediments will slowly release at a low concentration to supplement the pollutant concentration in the water, and then pollute the water. Therefore, from the perspective of water environment improvement, the sediment dredging measure is necessary and urgent for eliminating endogenous pollution, which is of great concern for its high technical requirements.

4.2.2 Environmental dredging of sediments

The sediment dredging is an engineering measure that hydraulically or mechanically excavates, transports and treats earthwork under water, with the major process systems covering sediment dredging system, sediment transport system, sediment dehydration and solidification system, sediment stacking yard system, auxiliary system, etc. Dredging can be divided into engineering dredging and environmental dredging based on the purposes. The engineering dredging is mainly for a certain engineering demand, such as for improving flood discharging capacity of rivers, improving navigation conditions and increasing storage capacity of lakes or reservoirs, while the environmental dredging is mainly to improve the water environment, with the purpose of removing pollutants contained in sediments by sediment dredging, so as to eliminate

Table 4.1 Comparison of three technologies

Technology	Advantages	Disadvantages	Cost effectiveness	Suitable conditions
Covering	Small disturbance without sediment movement; relatively stable chemical and hydraulic conditions after covering; some covering materials with the pollutant adsorption function; applicable to sediments with many kinds of pollutants, such as nutritive salts, heavy metals and POPs	The pollution problem will not be thoroughly solved because pollutants remain in the original places; the storage capacity will be reduced after the covering; it can't be implemented in large scale; it is not good for the biodiversity and it is vulnerable to strong water flows, stormy waves, etc.	Based on China's engineering experience in the river around Chaohu City, the construction cost is 98.5 yuan/m², including expenses for the transportation, covering construction, etc. and calculated on the basis of 50 cm-thick sands used for the covering	It needs to be implemented after the source control and is applicable to rivers after dredging or local areas of medium deep and deep-water lakes with easily available covering materials, but high dredging expenses and difficulty in finding stack yards; it is required that the covering layer will not be eroded by hydraulic conditions or stormy waves of lakes and can be supported by the bottom terrain
Passivation	Small disturbance without sediment movement; effectively reducing the suspension of sediments; applicable to sediments polluted by phosphorus or heavy metals	The ecological risk shall be considered due to the addition of chemical agents; the treatment efficiency may be different due to the nonuniform addition of chemical agents in situ; the treatment effects may be affected by environmental factors, such as water flow and temperature; the technical research is still immature, especially for pollutions except for NP; the passivation effects may be affected by the water flow or stormy waves	Based on overseas engineering experience, it only costs 1/5–2/3 of the dredging expense, and the passivation efficiency of pollutants reaches up to 50%–90%	It needs to be implemented after the source control and is applicable to locally heavily polluted areas of non-water source lakes or reservoirs; it is required that the passivation layer shall not be eroded site hydraulic conditions and stormy waves
Dredging	Increasing reservoir capacity; good effect due to complete elimination of endogenous pollution and ex situ disposal; applicable to the removal of various pollutants except for volatile ones; relatively mature technology	The ex situ stacking and disposal of sediments need long-term monitoring; it is difficult to eliminate secondary pollution caused by fine particles; benthic organisms are removed along with polluted sediments and the odor emitted during the dredging will have adverse effects on the surrounding environment	Based on the existing engineering experience of China, the comprehensive construction cost is about 30–50 yuan/m³, covering the dredging, stacking and disposal	It needs to be implemented after the source control and is applicable to lakes, heavily polluted river bends or estuaries with favorable stacking conditions and inexpensive land occupancy fees

endogenous pollution sources, reduce the release of sediment pollutants into water, improve the water environment of rivers, lakes or reservoirs, create internal conditions for further restoration of polluted water and also provide favorable environment for the restoration of water ecosystem. As for engineering dredging mainly for improving the discharging capacity or increasing storage capacity of lakes or reservoirs, it requires relatively simple dredging technology and is easily implemented. As for environmental dredging for improving the water environment, it has relatively high technical requirements.

At present, the sediment deposition is very serious in the backwater area of the Yangtze Three Gorges channel. Meanwhile, pollutants from surrounding water bodies of the Three Gorges Reservoir Region enter the Three Gorges Reservoir Region under the combined effects of point source pollution and non-point source pollution, then the pollutants enter the sediments as time goes on, resulting in the sediment deposition and pollution in the Three Gorges Reservoir Region. Sediments may release pollutants into water due to the sediment resuspension during traditional engineering dredging process, resulting in water eutrophication. Therefore, a reasonable and feasible environmental dredging construction technique is badly in need to reduce the sediment pollutant release into water, so as to achieve coexisting channel dredging and environmental benefits. Environmental dredging is a dredging method that combines the dredging in conventional construction with ecological restoration, soil and water conservation, resource utilization, environment renovation and other environmental protection contents in channel dredging engineering, which, by means of integrated treatment, finally achieves the dual goals of eliminating pollution sources in sediment segments of channels and creating favorable conditions for water ecological restoration. In order to achieve the said effects, in the environmental dredging process, corresponding design modifications shall be made for the dredging equipment, in addition to strictly prepared dredging schemes (usually covering the determination of dredging range, dredging depth, etc.).

The environmental modification generally means to modify conventional dredgers on the following aspects, and environmentally modified dredgers are the most widely used equipment in dredging engineering, which are all equipped with the differential global positioning system (DGPS).

1. Cutter suction dredger.
   It is mainly to modify conventional cutter heads to environmental cutter heads, and there are four types currently: ① environmental covered cutter; ② environmental vertical disk cutter; ③ environmental helical cutter; and ④ scraping suction head.
2. Bucket dredger.
   It is mainly to modify the bucket ladder. The upper part of the bucket ladder is enclosed and the buckets are equipped with exhaust valves so that the air can be expelled automatically from buckets in water, so as to eliminate the risk of causing turbidity.
3. Grab dredger.
   It is mainly to modify the grab bucket to an enclosed one to avoid the leakage of sludge during dredging.

4. Dipper dredger.

A movable guard is added to the general dipper bucket to enclose the sludge in the bucket, so that the sludge will not leak during the lifting of bucket.

Cutter suction dredgers and grab dredgers are common in practical engineering applications. For example, one Beaver 600 environmental cutter suction dredger, two modified general Beaver 1600 environmental cutter suction dredgers, one general modified domestic $120\,m^3/h$ cutter suction dredger and one modified domestic $120\,m^3/h$ environmental cutter suction dredger were used in the sediment dredging engineering of the Caohai in Dianchi Lake; the IHC Beaver 750 environmental cutter suction dredger made by the Netherlands was adopted in the sediment dredging engineering of the West Lake; the Water Classic cutter suction dredger made by Finland was adopted in the environmental dredging engineering of Wuhan Shuiguo Lake. Another example, in the sediment dredging engineering of the Suzhou River, the grab dredger's grab bucket was environmentally modified that welded steel plates were used to seal the gaps on both sides of the grab bucket and reduce the opening area of the bucket; and in the sediment dredging engineering of the Dongqian Lake, the $0.75\,m^3$ grab dredger was adopted for the lakeshore shallow-water area due to the plenty of rubbish and debris caused by frequent human activities.

After dredging, the release rate of nitrogen nutrient from sediments will gradually decrease with time. Research shows that the release rate of nitrogen nutrient has a significant difference within 1 hour after dredging with different types of cutter, and the nitrogen release rate is lower when a helical cutter is adopted than that when a general cutter is adopted, but such difference will be slightly 1 hour later. The decreasing trend of the release rate of nitrogen nutrient will gradually slow down over time and will become stable and approach zero after 5 hours. In terms of reducing the release of pollutants from sediments during dredging, helical cutters have certain advantages over general cutters. For different pollutants in sediments, the adoption of helical cutters can reduce the release of pollutants caused by dredging to a certain degree, which has a significant impact on the release of TN and COD, with the accumulative release amounts reduced by 14% and 15%, respectively, compared with those when general cutters are adopted and has a relative weak impact on the NH_3-N with the release amount reduced by 6% compared with that when general cutters are adopted. In fact, compared with sediments removed in engineering dredging, polluted sediments removed in environmental dredging are softer and easily cut and are also easily suspended and spread. The cutting by continuous linear blade of a helical cutter is relatively stable and has small impact on the sediment surface, which can minimize the sediment disturbance caused by cutter, while the general cutter sacrifices the environmental performance to pursue the cutting capacity and adaptive capacity, so the blade is discontinuous. Therefore, general cutters cause relatively large sediment disturbance during dredging, which facilitates the release of sediment pollutants.

Besides, some researches show that the accumulative release amount of sediment pollutants during dredging is the lowest when the aluminum salt is used as the passivator and an environmental covered cutter dredging equipment is adopted. The analysis shows that when an environmental covered cutter is adopted, the special helical cutter of an environmental cutter can reduce the contact area between sediments and cutter, so as to reduce the sediment disturbance and further reduce the sediment diffusion caused by dredging disturbance. Compared with environmental helical cutter, the general cutter has a larger sediment cutting face because there is an angle between the cutter and sediment surface. Due to the larger contact area between cutter and the sediment surface, the sediment disturbance becomes larger accordingly, which is not conducive to reducing the sediment diffusion. Moreover, the cover installed on general cutter has certain advantages in reducing sediment diffusion, primarily because when a general cutter is installed with a cover, the disturbance caused by cutter can only be reduced to a certain degree due to the smaller cutting area between the cutter and sediments when the cutter bottom cuts into sediments. During dredging, the environmental covered cutter shows certain superiority under the combined effects of the absorption and stabilization of sediment pollutants by the passivator polyaluminum chloride (PAC) and the reduced cutting area between the cutter and sediments due to the cutter cover.

During sediment dredging, the sediment diffusion caused by dredging disturbance is the primary reason for the increase in accumulative release amount of sediment pollutants. Therefore, the reduction of dredging disturbance is the main means to reduce the release of sediment pollutants. As for dredging with general cutter, the sediment disturbance intensity is directly affected by the cutter's geometric dimension and shape, while the special helical cutter of environment cutter minimizes the sediment disturbance intensity. Besides, the releasable amount of sediment pollutants is reduced because of the obvious early passivation effect of the passivator. Therefore, the covered helical cutter has dual effects on reducing the sediment disturbance in the environmental dredging of COD pollutant. Meanwhile, the use of passivator also has a certain inhibiting effect on the release of sediment pollutants in the environmental dredging. Analyzed from the aspect of passivator, the hydroxyapatite $[Ca(OH)(PO_4)^3]$ precipitation formed by the $H_2PO_4^-$ from calcium nitrate will absorb and precipitate pollutants during the dredging; or under the anaerobic environment inside sediments caused by the passivation mechanism of calcium nitrate, microorganisms replace the role of oxygen in water and oxidize organics with nitrate as the electron acceptor and calcium nitrate passivator as the oxygen acceptor, so as to reduce the TP content. The environmental covered cutter shows certain superiority under the combined effects of the calcium nitrate's absorption and stabilization of sediment pollutants due to the passivator and the reduced cutting area between the cutter and sediments due to the cutter cover. As for the difference in influence on the release amount of NH_3-N from sediments, the electron acceptors in sediments increase after calcium nitrate enters water, thereby speeding up the metabolism of organic nitrogen, resulting in more NH_3-N released into water. However, some calcium hydroxide flocs in sediments formed by calcium nitrate in water have a certain apportion effect on NH3-N, and the inhibiting effect is greater than the effect on promoting the metabolism of organic nitrogen.

The dredging of polluted sediments plays an irreplaceable role in improving water environment, so we need to analyze and evaluate the results from the dredging engineering, and the sediment dredging brings multiple environmental benefits:

4.2.2.1 Elimination of endogenous pollution

After dredging, the content of organics, nitrogen, phosphorus and heavy metals in sediments will gradually drop, which can alleviate the water eutrophication to a great extent and eliminate the impact of heavy metals on water, indicating that the sediment dredging is an effective means of reducing sediment endogenous load.

4.2.2.2 Improvement of water quality

The removal of polluted sediments not only removes pollutants accumulated in sediments for one time but also reduces the consumption of dissolved oxygen in water as well as the pollutant exchange between sediments and water, thereby improving the water quality. Therefore, water quality indicators such as the water transparency, SS, BOD and NH3-N will also be changed before and after the dredging.

4.2.2.3 Other eco-environmental effects

The sediment dredging can promote the aquatic eco-environmental restoration to a certain degree. The dredging engineering removes polluted sediments to provide basic survival conditions for the restoration of aquatic macrophytes and contributes to the ecosystem restoration. Meanwhile, the growth of algae and some invasive alien species will also be reduced due to the improvement of water quality, which will promote the restoration of aquatic ecosystem. Besides, the sediment dredging engineering can increase the reservoir capacity and corresponding storage capacity, which helps relieve the imbalance between supply and demand of water resources. Moreover, the improvement of water quality can improve the water environment functions and promote the exploitation and utilization of water resources. At the same time, the waterfront land reclamation and ecological restoration of storage areas can obviously improve the landscape and ecological environment of dredged areas and surrounding areas.

Although the dredging has advantages such as quick effect, elimination of endogenous pollution and improvement of water quality, it also has some negative effects from a long-term perspective:

1. Impacts on water environment.

    If inappropriate dredging schemes are adopted, such as schemes with improperly selected dredging area, dredging depth and other parameters, or ineffective technical measures, pollutants in sediment pore water are very likely to re-enter the water, and the pollutants released may enter the surface water under the effect of water flow and wind. In recent years, sediment dredging has been carried out for the treatment of water eutrophication in many areas. However, the sediment dredging in some areas didn't achieve expected results and even destroyed balance of nutrients such as nitrogen and phosphorus in water, resulting in further deterioration of water quality and extremely frequent algal blooms. Besides, the discharge of residual water from storage areas will also cause impacts on the water environment. Because agents are added during the residual water treatment to promote the coagulation, the type and amount of the coagulant may cause adverse impacts on the receiving water.

2. Impacts on ecological environment.
   The sediment dredging may cause damages to land plants and animals. For example, the land occupancy and earth excavation for dredging engineering may destroy vegetation in construction areas and also cause changes in the land utilization purposes and structures and then affect the habits of animals and plants. Such impact is unrecoverable in the short time, but it is temporary and such destruction will be alleviated after the construction.

The sediment dredging will also cause adverse impacts on aquatic life that benthic animals will be more or less removed during the dredging process, resulting in damage to the original benthic ecosystem. After the dredging is completed and before a new benthic ecosystem is built, the river ecosystem will be fragile and vulnerable to algal blooms. For instance, there was large-scale outbreaks of algal blooms after dredging of the Xuanwu Lake in Nanjing in the early 21st century. In summary, the impacts of sediment dredging engineering on aquatic ecological environment are short-term and temporary, which will be gradually alleviated after the construction.

To sum up, the adverse impacts caused by sediment dredging engineering are controllable and restorable. Therefore, we need to take proper control measures and develop reasonable dredging schemes in advance during the construction and take restoration measures according to local conditions after the dredging, so as to realize environmental dredging in the real sense.

4.2.3 Disposal and reutilization of sediments and residual water

4.2.3.1 Disposal and reutilization of sediments

First, the areas for sediment placing, namely the common sediment dumping area, collection area and landfill, shall be properly designed and their site selection shall meet the following principles:

1. Meet relevant planning requirements
2. Meet environmental requirements and cause no secondary pollution to nearby soil and water
3. Storage areas shall have good anti-seepage measures and residual water discharge channels
4. Storage areas shall be selected on the principle of proximity, so as to reduce the sediment transportation cost and impacts caused during the transportation
5. Try to minimize the farmland occupancy and vegetation destruction to reduce impacts on the ecological environment. The selection rationality of storage areas shall be analyzed according to the above principles in the environmental assessment.

In practice, storage areas shall be designed according to the engineering demands on the principles of adaptation to local conditions and use of local materials. For example, in the sediment disposal of Caohai in the Dianchi Lake, the geomembrane was used inside the cofferdams of storage areas, and the peat layer widely existing underground in areas around the lake was used at the bottom of storage areas as pollutant adsorption layer. Usually, sediments will be first subject to dehydration treatment and modification treatment with flocculant and stabilizer as modifiers, then many different methods are widely adopted in China for the recycling and comprehensive utilization of treated sediments. For example, sediments of the Shanmei Reservoir formed mud cakes after a series of treatment and then were transported to a brickyard for brick making; $153 \times 103\,\text{m}^3$ lake sediments dredged in Massachusetts, USA, were used as soil conditioner; the sediments dredged from New York and New Jersey ports were used as soil and raw materials for cement production; sediments of the Wenruitang River were piled up in storage areas, mainly used as part of the "waterfront land reclamation" project.

4.2.3.2 Disposal and reutilization of residual water

After the natural sedimentation of polluted sediments, the excessive water discharged from storage areas is called residual water, which contains a lot of nitrogen, phosphorus, organics, heavy metals and other pollutants that can be attached to fine particles and then suspended in residual water. In general, the methods for residual water treatment are divided into physical, chemical and biological methods.

Physical methods such as filtration are usually to set some filters in residual water treatment tanks. For example, in dredging engineering of the Yuhu Lake in Wuhan, S-shaped partition ridges were set in the sedimentation tank for further sedimentation of residual water, and two three-layer filters (first layer: large-size cobblestones; second layer: coarse sands; third layer: fine sands) were set for the filtration to make the residual water meet discharge standards.

As for chemical methods, the flocculation is the most common method for residual water treatment at home and abroad, which has the advantages of no need for external power, easy operability, good adaptability to changes in water quality and quantity, diverse and cheap agents, small treatment area occupied, simple facilities, low costs and convenient construction. According to relevant standards, the residual water is allowed to be discharged only in case of the treatment rate >90%, SS mass concentration <150–200 mg/L, or meeting discharge standards after treated through small purification treatment facilities, oxidation pond, etc. Usually, the treated residual water with environmental indicators close to those of natural water can be discharged back into lakes to realize the recycling of water resources.

At present, there is also an ultra-magnetic purification treatment technology in the black and odorous water treatment, which is especially suitable for removing light impurities such as fine SS and TP that are difficult to be settled, so as to ensure no secondary pollution to the environment caused by residual water discharged back into rivers.

4.2.4 Water circulating measures

The core of water supplement, water circulating and hydrodynamic maintenance measures is to divert water from other rivers to dilute pollutants and restore natural river courses, with main methods as follows: (1) Clean surface water can be diverted to supplement water bodies under treatment and promote the pollutant transport and diffusion, so as to improve the water quality. For example, the scientific management and utilization of urban stormwater not only can supplement the water volume but also can improve water fluidity, accelerate water replacement and improve water quality. This method is applicable to the long-term water quality conservation of stagnant polluted water bodies, semi-closed and closed polluted water bodies. (2) Treated urban sewage satisfying water quality requirements for reclaimed water can be discharged into treated urban water bodies to increase the water flow and reduce hydraulic retention time. Urban reclaimed water can be fully used as supplemental water source of urban water bodies to increase the ecological water of rivers and lakes. As a stable non-conventional water source for cities and towns, the reclaimed water is an economically feasible supplemental water source with great potential and should be preferably utilized. This method is applicable to the long-term water quality conservation of water-deficient cities or treated polluted water bodies in withered water period. (3) Engineering measures can be taken to increase the water flow velocity, so as to improve the reaeration capacity, self-purification capacity and water quality of water bodies. The reasonable and effective interconnection of urban river and lake systems can be strengthened to build urban water systems of benign cycle. This method is applicable to closed water bodies with slow water flow velocity.

4.2.4.1 Bypass purification and water circulating technology

The water circulating technology is to increase the water flow velocity by engineering measures to improve the reaeration capacity, self-purification capacity and water quality of water bodies, which is applicable to closed water bodies with slow water flow velocity.

Without increase in water demand, the artificially increased water circulation dynamic can also effectively improve the water fluidity, self-purification capacity of water bodies and the quality of water environment. Many studies have also shown that good hydrodynamic conditions help to alleviate the eutrophication situation of slow-flow water bodies. Moreover, the hydrodynamic circulation is in line with the natural ecological principle than methods such as disinfection and algae killing and flocculation.

The hydrodynamic control of black and odorous water can reduce the pollutant concentration and water stratification of loaded branch and increase the DO concentration in water. The measure is divided into two types, namely, hydraulic regulation method and mechanical regulation method, and the hydraulic regulation method is mainly the water circulating method for black and odorous water. The water circulating method can adjust the hydraulic retention time and improve hydrodynamic conditions and self-purification capacity of water bodies. The another method is to create the water circulating convection in a relatively closed water by the operation of submersible plug-flow mechanical equipment to realize slow and uniform flow of water, thereby improving the water flow patterns. It can also realize the interconnection among multiple water systems in a larger system to facilitate the water exchange. The water circulating technology is an indispensable measure for the long-term improvement and conservation of water quality.

First of all, current situations of water systems in a study area shall be investigated, and the water system planning layout in the study area shall be preliminarily determined by following the water system planning principles of the corresponding area and with overall consideration of multiple planning requirements of the study area. Second, the multi-natural patterns river improvement method shall be adopted according to the geographical characteristics of the study area and with consideration of the water system landscape and ecological results to build ecological rivers; the existing water system framework and river terrain shall be fully utilized to divert clean water to cities by river improvement and structure construction, and the water shall be timely regulated and stored for flood control and waterlogging drainage, giving play to the roles of both landscape and practical application; the sewage shall be collected and discharged into ecological wetlands after meeting discharge standards for further purification; the location and requirements of hydraulic structures and the river layout shall be determined reasonably to enable the planned water system to protect, exploit and utilize existing water system, regulate and control flood and sediments, facilitate water conservancy and prevent flood disasters and create an integrated ecological water environment; engineering measures shall be adopted to restore interrupted river courses to make them connected to other river courses; river courses in cities shall be connected with control structures established at junctions to reasonably adjust the water level and flow.

On the principle of natural ecological purification of water bodies, the water circulating convection created in a relatively closed water by the operation of submersible plug-flow mechanical equipment can improve the water flow patterns. In a large system, a dynamic force can be applied to the water body based on the basic principle of hydrodynamics to connect multiple water systems and turn the static water body into a dynamic one, so as to realize slow and uniform flow of water, facilitate the water exchange, increase the DO content in water, destroy the living environment of microcystis, realize the algae reproduction suppression and activate the water body's purification function.

The characteristics of water circulating technology are as follows:

1. Applicable scope: it is applicable to the pollutant treatment and water quality conservation of urban slow-flow rivers or pit-pond areas, which can effectively improve the water liquidity.
2. Advantages and disadvantages of the technology.

   a. Advantages.

      ① Water systems can be reasonably connected by lift pumps, and the flowing of water can be realized by making use of wind or solar energy. In non-rainy seasons, stormwater pump stations or storm sewers near water bodies can be used as return water systems, and attention shall be paid to the setting of circulating water outlets to reduce the river bed scour or lakebed by circulating water.

② By the connection, planning, treatment and control of water systems, it can realize flood disaster prevention, water conservancy facilitation, waterlogging drainage, water utilization and shipping transportation.

③ It can realize the ecological balance of water environment, enhance the self-purification capacity of water systems and reduce the degree of water pollution.

④ The urban water systems can be closely combined with urban green space systems, forming important parts of the overall urban spatial pattern.

⑤ The water system treatment can treat pollution sources along rivers and protect the water quality of rivers to ensure the water quality safety of urban water sources and also improve the city image and enhance the city grade.

b. Disadvantages.

① The water system planning focuses on the river flood control and water resource allocation but neglects the ecological benefits.

② The water system planning usually aims for flood control, waterlogging drainage, shipping and production and rarely focuses on shaping the urban landscape image by using water systems.

③ The water system planning shows insufficient understanding of the function of urban water systems as important recreation and daily activity space of urban residents.

④ The water system planning fails to make full use of water systems as important corridors of cities and give play to their all kinds of value.

⑤ Some projects need the laying of water channels, resulting in relatively high construction and operation costs, high implementation difficulty and continuous operation maintenance; and the ecological risk assessment shall be conducted on the connection of river and lake systems to avoid blindness.

4.2.4.2 Ecological water supplement technology

Due to the irregular shapes and self-closure, urban black and odorous waters are characterized by slow flow, poor self-purification capacity, local dead water zone and pollutant accumulation, finally causing eutrophication or even algal blooms. Some urban black and odorous water bodies are water-deficient (dried up), and their water areas are greatly reduced and habitats are degraded due to the insecure ecological basic flow and lack of necessary water volume of water ecosystem. Therefore, the ecological water supplement is an important measure for the prevention and treatment of black and odorous water because it can (1) supplement the ecological basic flows of rivers and other water bodies to ensure the ecological environment water and (2) greatly improve the environmental capacity and water resource sustainable utilization capacity.

As for water bodies with small or little ecological basic flows, the ecological water supplement can be adopted to supplement the water with clean water, reclaimed water, stormwater, etc. As for stagnant water or slow-flow areas, it is encouraged to adopt technologies such as internal or external circulation. At present, the water supplement technology mainly includes two modes, namely, supplement with clean water and supplement with reclaimed water, and the latter one is the main mode for the water supplement of urban water bodies.

1. Supplement with clean water.

   The supplement with clean water means to divert clean surface water to supplement water bodies under treatment and promote the pollutant transport and diffusion, so as to improve the water quality, which is applicable to the long-term water quality conservation of stagnant polluted water bodies, semi-closed and closed polluted water bodies. Besides, the exploitation of and supplement with clean surface water can increase the environmental capacity of water bodies, but attention shall be paid to the dynamic balance of water to avoid affecting or destroying the functions of water bodies nearby.

2. Supplement with reclaimed water.

   Treated urban sewage satisfying water quality requirements for reclaimed water can be discharged into treated urban water bodies to increase the water flow and reduce hydraulic retention time. As a stable non-conventional water source for cities and towns, the reclaimed water is an economically feasible supplemental water source with great potential and should be preferably utilized. It is applicable to the long-term water quality conservation of water-deficient cities or treated polluted water bodies in withered water period. However, the supplement with reclaimed water always requires the laying of pipes, intensive monitoring on the quality of supplemental water and additional investment in appropriate advanced purification measures taken for supplemental reclaimed water.

Moreover, the ecological water supplement can prevent running water from being putrid, and the most important role of water circulating flow is to create favorable conditions for ecological purification, which can make pollutants degraded by efficient microorganisms, supplemental nutrients in water fully diffused everywhere in water bodies and play their role and also contributes to the water reaeration. During the water circulation, the residual organics, inorganic nutrients, algae, SS, etc., in the water that has been purified by water treatment systems such as biological filter or sand filter will be further removed by enhanced biological or physical process. Due to their low removal rate of high-load pollutants, these measures can be taken as auxiliary measures for the water purification. When an ecosystem is seriously impacted or a water ecosystem is unable to be self-restored in case of lots of pollutants entering water abruptly, appropriate amount of clean water can be supplemented to reduce the pollutant concentration. Besides, the supplement with clean water can also make up the loss of water entering lakes. The best method for water supplement is to pump and use landscape water for green landscape irrigation and then supplement with clean water, thereby achieving the purposes of water flowing and water quality improvement.

The characteristics of ecological water supplement technology are as follows:

1. Applicable scope: it is applicable to the water supplement of urban water-deficient water bodies and the hydrodynamic improvement of stagnant or slow-flow water bodies, which can effectively improve the water fluidity.

2. Advantages and disadvantages of ecological water supplement technology.

   a. Advantages.

   ① It utilizes the urban reclaimed water, urban stormwater and clean surface water as the supplemental water sources of urban water bodies, which can increase the water fluidity and environmental capacity, give full play to the role of sponge city construction and enhance the retention, storage and purification of urban rainfall runoff.

② The exploitation of and supplement with clean surface water can increase the environmental capacity of water bodies.

b. Disadvantages.

① The supplement with reclaimed water always requires the laying of pipes.
② The water supplement expense sharing mechanism shall be specified.
③ It is not advocated to implement the ecological water supplement by way of long-distance water diversion.
④ The supplement with reclaimed water shall be adopted with appropriate advanced purification measures to meet the quality requirements for supplemental water.

4.2.4.3 Regulation of basin water resources

Water resources in basins are mainly regulated by water conservancy projects such as reservoirs. In the past, the flood control and beneficial utilization scheduling were the primary considerations for the reservoir scheduling, while now the drought relief, water supply, reservoir deposition reduction, ecological environment and emergency scheduling in response to water pollution emergencies shall also be taken into consideration. The water diversion technology means to divert stagnant water stored in rivers "from main streams to tributaries, downstream to upstream, river courses to river banks, river courses to wetland and inside rivers" by following the concept of "circular water". The repeated "five cycles" can facilitate the water flowing, which not only solves the water shortage problem of the upstream river channels but also facilitates the running of stagnant water and improves the water quality, giving a new concept of "circular water" in the practice of river treatment and management. The water diversion technology not only can dilute the pollutant concentration in black and odorous water with a large amount of clean water but also can promote the pollutant diffusion, purification and output. It has obvious effects on the treatment of urban black and odorous water bodies with high pollution load, insufficient hydrodynamics and low environmental capacity but is also a waste of water resources, so non-conventional water sources such as reclaimed water and stormwater resources should be used as much as possible.

As an important means of water quality control for river and pond water systems, the water diversion refers to the water resource utilization method that makes full use of external clean water resources to improve hydrodynamic conditions and increase environmental capacity of water bodies by reasonable scheduling of engineering facilities such as gates and dams, so as to improve the water quality.

After being seriously polluted, a water body may become stagnant water and completely loses its self-purification capacity. The water transfer and diversion, on the one hand, can increase the DO in polluted water, and on the other hand, can maintain the flowing state of water, increase the DO concentration in water and enhance the self-purification capacity of water.

Advantages and disadvantages of the water diversion technology:

1. Advantages.

   ① After the water circulating flow is improved, it can prevent the algal blooms and also solve the upstream water shortage problem and improve the water environment of river courses.
   ② It can improve the water quality and also promote the formation of ecological and natural landscape in surroundings.
   ③ More comfortable ecological river courses can be built to realize the human–water harmony.
   ④ The water body remediation pond, after being built, can effectively remove the nitrogen, phosphorus and other nutrients in water, increase the DO in water and improve the self-purification capacity of water to create a virtuous water circulation system.
   ⑤ It can enhance the carrying capacity of water environment.

2. Disadvantages.

   ① The poor water regulation is likely to cause water quality decline and water environment deterioration.
   ② Once the water environment carrying capacity of the water diversion project is exceeded, it will be more difficult to treat the river and harder to restore the balance of water environment.

The Three Gorges Reservoir is a seasonally regulated reservoir with a regulation storage capacity of 16.5 billion m^3 and a flood control capacity of 22.15 billion m^3. Located at the junction of the middle and upper reaches, it not only plays an important role in the flood control of the middle and lower reaches but also increases the main stream runoff in the middle and lower reaches from every December to next May, which raises the water level in each river segment to varying degrees, improves the domestic and production water intake conditions and navigation conditions in the middle and upper reaches of Yangtze River during withered water period and can alleviate the drought in the middle and upper reaches of Yangtze River in withered water period and the saltwater intrusion at Yangtze River estuary.

Main effects of the Three Gorges Reservoir's water supplement to the middle and lower reaches: (1) It helps the realization of the shipping, ecological and drought-relief functions of the middle and lower reaches. (2) In withered water period, it supplements water to the lower reaches combined with the power generation function, with the outflow increased by 1,000–2,000 m^3/s compared with the inflow, effectively alleviating the low water level situation of the downstream river in withered water period. (3) It increases the navigation channel depth by 0.72 m on average, greatly improving the navigation capacity of the middle and upper reaches of Yangtze River (especially the segment from Chongqing to Wuhan). (4) The operation of the Three Gorges Reservoir increases the inflow to Yangtze River estuary in withered water period, which can effectively suppress saltwater intrusion at the estuary. In sum, except for the storage period at the end of flood season, it evenly supplements water to the middle and lower reaches in withered water period, increasing the discharged flow.

The basin water resource scheduling includes not only water supplement scheduling but also ecological scheduling. The ecological scheduling can be interpreted in a broad sense or a narrow sense. In the broad sense, it includes two aspects: ecology and environment, such as the scheduling for reduction of deposition at the end of the Three Gorges Reservoir Region, control of algal blooms at reservoir bends, water and sand supplement to the lower reaches and estuary, rebuilding of aquatic life habitats and spawning of flagship species; while in the narrow sense, it is mainly for the spawning of specific protected species (such as the four major Chinese carps). Combing with the conditions of inflow from the upstream and taking advantage of the opportunity of accelerated drawdown before flood season, the Three Gorges Reservoir changes the water discharge process to artificially create the flood peak process that has the necessary hydrological and hydraulic conditions suitable for the spawning and reproduction of the "four major Chinese carps". With five times of pilot ecological scheduling, the flow fluctuation process lasting 3–10 days, with the flow increase of 1,000–6,000 m^3/s, was continuously made during the spawning period of the four major Chinese carps through regulation and storage of the Three Gorges Reservoir. At the same time, it should be guaranteed that the flow downstream of the Three Gorges Reservoir was obviously increased and the main stream water level had a certain variation.

4.2.5 Artificial oxygenic aeration technology

The black and odorous water is a common phenomenon in urban river networks of China, the root cause of which is the lack of DO in river water. The DO in river water mainly comes from the atmospheric reaeration and photosynthesis of aquatic plants, and the former one is the main source of the DO in water. The river water oxygen consumption mainly includes the oxygen consumption by reducing substances, biochemical oxygen consumption by dissolved and colloidal easily degradable organic pollutants, oxygen consumption by NH_3-N nitration and the oxygen consumption by solid organic pollutants and refractory organics such as river sediments. When the river reaeration rate is greater than the oxygen consumption rate, the organics will be decomposed by aerobes, causing the decline in DO content in water and the DO concentration lower than the saturation value; then the oxygen in water surface atmosphere will be dissolved in the river water to supplement the consumed oxygen to make the DO in river water gradually back to normal. Such process is called the river self-purification.

The DO content is an important indicator reflecting the water pollution condition. The change process of the DO concentration in polluted water reflects the river's self-purification process. When the river reaeration rate is smaller than the oxygen consumption rate and the organic content in water and river sediments is too high, the DO is consumed too fast to be supplemented by the atmospheric reaeration and algae photosynthesis, then the DO in water will gradually decline and even be used up, resulting in organic aerobic decomposition turned to anaerobic decomposition, incomplete decomposition of organics and release of a large amount of toxic substances. It will cause serious damage to the water ecosystem, deterioration of water quality and black and odorous water that can't be self-restored simply by the natural aeration effects (atmospheric reaeration and algae photosynthesis). Therefore, the artificial oxygenic aeration technology shall be adopted to increase the oxygen transfer and diffusion.

Main effects of the artificial oxygenic aeration technology are as follows: (1) It can quickly oxidize blackness-causing and odor-causing matters such as H_2S, methanethiol and FeS to effectively alleviate the black and odorous water situation; (2) it can improve the water quality, because some harmful organics will be gradually degraded to low molecular organics or inorganics harmless to human body after the DO in water is increased; and (3) it can restore the ecological balance. When the water of a river or lake is lack of oxygen, fish and shrimp and other aquatic life may die or even become extinct, but the oxygenic aeration can facilitate the degradation of massive harmful pollutants and increase the DO necessary for the aquatic life, so as to restore the river or lake's ecological balance.

4.2.5.1 Introduction to traditional oxygenic aeration technology

The traditional oxygenic aeration technology is to supplement oxygen regularly or irregularly by aeration devices installed in water to maintain the aerobic conditions between the sediment interface and water. Under normal circumstances, the phosphorus is deposited in sediments but will be released from the sediments under anaerobic conditions. Therefore, the traditional oxygenic aeration technology is applicable to the treatment of phosphorus-caused water eutrophication. The water aeration can effectively suppress the release of phosphorus from sediments and has a certain effect on controlling the stink and algal blooms in rivers and lakes.

Since the 1940s, the oxygenic aeration technology has been widely applied to the treatment of urban rivers and lakes at home and abroad. Foreign researchers applied the oxygenic aeration technology for the treatment of the Flambeau River, Mississippi River, etc., and all achieved good results. The aeration can significantly increase the DO content in water to restore the water habitat, control organic pollutants and eliminate the black and odorous water. As the first country that treated polluted rivers by the pure oxygen aeration method, Germany treated the Emshe River, etc. with the pure oxygen aeration system in the 1990s, which also indicated that the oxygenic aeration technology was more economical than other technologies. China began to use the aeration technology for the treatment of polluted rivers and lakes in the 1990s. During the Asian Games in 1990, the aeration equipment was used to oxygenate a river segment with a length about 4 km in Qinghe, Beijing, and the black and odorous water phenomenon was significantly improved 47 days later, with obviously improved water transparency, the DO content at the aeration site increased from 0 to 5–6 mg/L, and DO content downstream the aeration site increased by 2–3 mg/L on average. Shanghai Research Academy of Environmental Sciences conducted an oxygenic aeration experiment on Suzhou River in 1998 and the experiment result showed that: the main water quality indicators were restored to the meet the national Class IV standards; various pollutants were effectively removed; and the water eco-environment was gradually improved. Multi-functional purification ships were used to conduct an aeration restoration experiment on Zhangjiabang River segment in Pudong New Area, Shanghai.

After one and a half of treatment, the water quality indicators were significantly improved and basically met the water quality standards for Class IV surface water. These studies have fully shown that the oxygenic aeration technology is an effective water treatment technology.

4.2.5.2 Classification of traditional oxygenic aeration patterns

Aeration facilities of traditional oxygenic aeration technology mainly include blower–microporous air distribution tube aeration system, pure oxygen–microporous air distribution equipment aeration system, impeller suction plug-flow aerator, underwater jet aerator, etc., and each type has its advantages and disadvantages. Waterwheel aerators, impeller aerators and jet aerators of different aeration modes were selected to conduct field experiments on black and odorous rivers. The results have shown that in terms of the aeration efficiency, waterwheel aerators are the best, followed by impeller aerators and the jet aerators are the worst. Especially, waterwheel aerators can quickly start the bacteria–algae ecosystem of a river and form a relatively long visible clean aerobic green river segment in the downstream, forming the unique landscape of "black in the upstream, green in the middle and black in the downstream" of the treated river. If impeller aerators are used, such phenomenon can also be observed in the downstream of the aerators, but the green river segment is shorter than that downstream of waterwheel aerators, while such phenomenon can't be observed if jet aerators are used.

Both waterwheel aerators and impeller aerators form the gas–liquid contact interfaces by generating hydraulic jumps of liquid-phase fluid. The kinetic energy acts on the motion of heavy liquid-phase fluid and the light gas-phase fluid is passively contacted to form locally continuous gas–liquid contact interfaces around the impeller or rotary brush (wheel) stirring area. Therefore, they have the saponification that can separate grease, purify water, increase DO and will not stir up impurities such as organic debris that have been deposited in sediments. They are also characterized by powerful current, high DO content and small noise. Characterized by very fast rotary speed and primarily vertical stirring, impeller aerators have a certain stirring impact on river sediments, which will increase the impurity concentration in water and affect the water quality; waterwheel aerators mainly generate horizontal plug flow that has less stirring impact on river sediments and high aeration efficiency; jet aerators suction liquid-phase fluid by jetting liquid-phase fluid to form gas–liquid contact interfaces, which have strong mass transfer capacity and stirring cutting effect but greatly stir the river sediments because the jet nozzles are located at the river bottom. The aeration effect is mainly for the oxidation and biochemical decomposition of reducing substances in the river sediment pollutants and is not suitable for the oxygenic aeration of small and medium-sized black and odorous rivers. Characteristics of traditional oxygenic aeration technology are as follows:

1. Applicable scope.
   The technology is applicable to the treatment of lake water eutrophication caused by increased TP content, improvement of urban black and odorous water, increasing the DO in water, suppressing the anaerobe breeding, etc.

2. Advantages and disadvantages of traditional oxygenic aeration technology.

    ① Advantages: high controllability, supporting selective starting according to water conditions; small land occupancy area, not requiring a large number of structures and equipment floating on water surface; small investment in equipment; no special requirements for operators, only requiring simple training.

    ② Disadvantages: poor treatment effect on flowing water, applicable to static water; weak capacity for TN removal, unobvious effect on treatment of eutrophication caused by nitrogen and phosphorus and limited effect on the eco-environment restoration.

4.3 Countermeasures for integrated water environment treatment

4.3.1 Countermeasures for black and odorous water treatment

The black and odorous water treatment is a complicated systematic engineering. In a city, the urban water body is an essential important part of the urban ecosystem, which has multiple functions, such as water circulation and improving the urban climate. Due to the rapid development of cities in China, the cities' water environment quality is under serious threat. Most of urban water bodies are polluted, which have seriously affected the living quality of residents. Therefore, we should not ignore the analysis on causes of black and odorous water and measures for long-term maintenance after treatment.

The treatment of black and odorous water shall be considered in the long run, and the treatment schemes shall be scientifically developed according to the technical route of "exogenous pollution reduction, endogenous pollution control, water quality purification, water supplementing and circulating and ecological restoration".

4.3.1.1 Relevant requirements in China

1. Water pollution control action plan.
   In April 2015, the *Notice of the State Council on Issuing Water Pollution Control Action Plan* (GF [2015] No. 17) was issued, commonly known as "Ten Measures for Water Pollution Control", wherein it is clearly specified that:

   > the people's governments of cities shall be the responsibility subjects for the treatment of black and odorous water bodies, and the MOHURD shall take the lead to guide the local implementation and propose goals together with ministries and commissions like the Ministry of Environmental Protection, Ministry of Water Resources and Ministry of Agriculture: by the end of 2017, cities at the prefecture level or above shall realize no large area of flotsam on river surface, no waste on riverbanks and no illegal sewage outfall, and the black and odorous water bodies shall be basically eliminated in the built-up areas of municipalities directly under the central government, provincial capitals and cities specifically designated in the state plan; by the end of 2020, the black and odorous water bodies shall be controlled within 10% in urban built-up areas of cities at the prefecture level or above; by 2030, the black and

odorous water bodies shall be generally eliminated in the built-up areas of all cities in China.

2. Guideline for urban black and odorous water treatment.

   In August 2015, the *Notice of the MOHURD and Ministry of Environmental Protection on Issuing Guideline for Urban Black and Odorous Water Treatment* (JC [2015] No. 130) was issued, which clearly specified that cities at the prefecture level or above should publicize the local black and odorous water treatment plans (including the names of black and odorous water bodies, persons in charge, treatment completion deadlines, etc.) by the end of 2015 and accept the public supervision. Together with the said Notice, the *Guideline for Urban Black and Odorous Water Treatment* was also issued, mainly covering the definition, identification and grading of black and odorous water body, treatment scheme preparation, treatment technology, treatment result evaluation, organization implementation and policy guarantee, aimed at guiding districts concerned to accelerate the black and odorous water treatment, improve urban ecological environment and promote the urban ecological civilization construction. Thereafter, the MOHURD organized the preparation of *Urban Black and Odorous Water Treatment—Technical Guide for Treatment of Drainage Outlets, Pipelines and Inspection Wells (Trial)*, which put forward requirements for the investigation and treatment of drainage outlets, inspection and evaluation on drainage pipes and inspection wells, pipeline restoration and treatment, pollutant interception, regulation, storage, in situ treatment, maintenance management of drainage pipes, inspection wells and drainage outlets, etc.

3. Evaluation on black and odorous water treatment effects.

   In May 2017, the *Notice of the MOHURD and Ministry of Environmental Protection on Conducting Evaluation on Urban Black and Odorous Water Treatment Effects* (JBCH [2017] No. 249) was issued, in which the two ministries detailed the requirements for the evaluation on black and odorous water treatment effects and explicitly stipulated that:

   > municipalities directly under the central government, provincial capitals and cities specifically designated in the state plan should achieve initial success in the black and odorous water treatment by the end of 2017 and realize the goal of long-term clear water in 2018; other cities at the prefecture level or above should achieve initial success in the black and odorous water treatment by the end of 2019 and realize the goal of long-term clear water in 2020.

4.3.1.2 Treatment measures in China

In order to implement the *Water Pollution Control Action Plan* of the State Council and guarantee the long-term clear water of rivers, all Chinese cities promote the work mainly from two aspects, namely, deepening working mechanism and improving engineering measures, mainly including the deepened working mechanism mode centered on "implementing 'river chief system', strengthening the collaboration mechanism among departments and regions, tightening supervision mechanism and improving guarantee mechanism" and engineering measures focusing on "strictly controlling pollution sources, increasing water circulation, carrying out ecological restoration and proposing the sponge city concept". Among them, the main measures are how to strengthen the mechanism, how to correctly identify pollution sources and how to maintain the water state. According to statistics, most cities have achieved remarkable results in improving the collaboration mechanism among departments, monitoring water quality and carrying out ecological restoration; and in order to achieve the long-term clear water, a few cities have carried out different forms of cross-regional treatment, evaluated and fed back the effects of innovative treatment measures and proposed to build "sponge cities" to promote the realization of long-term clear water.

The collaboration mechanism consisting of inter-departmental collaboration and inter-regional collaboration helps the departments' labor division internally and the collaboration externally. The water conservancy bureau, together with multiple departments such as local environmental protection bureau, land and resources bureau shall participate in and set up the ecological environment construction headquarters to monitor the water environment in real time.

An important step of the black and odorous water treatment is to analyze the cause and identify the source. The water quality monitoring is a process that monitors and measures the types, concentrations and variation trends of various pollutants in water to evaluate the water quality.

Routine patrols can timely remove flotsam on rivers and lakes and reduce shoreside waste to reduce visual objects' pollution to water and also can timely find and deal with new sewage outfalls to rivers. In the aspect of intensifying routine patrols, the solution is mainly to set up patrols to periodically patrol key rivers and lakes.

Public supervision not only can engage residents in the treatment to realize real public management but also can improve the self-consciousness of administrative authorities to a certain degree to make them do the pollution treatment work well consciously. In order to better implement the public supervision, online announcement or offline special issue or column is adopted to let the public know in real time the river and lake treatment process and various problems later.

The black and odorous water treatment is just the first step, and the long-term clear water also needs the improvement of guarantee mechanisms, including the effect evaluation mechanism and fund guarantee mechanism. The effect evaluation means to make an objective and effective evaluation on the current situation after the water environment treatment and improvement through investigating the water environment quality after treatment and satisfaction of residents nearby by such means as public review and monitoring on physical and chemical indicators of water, so as to improve the urban living environment. In addition, the Internet can also be used for feedback on the pollution treatment situation.

4.3.1.3 Relevant requirements in Chongqing

1. Department functions.
   Chongqing Municipal People's Government has successively issued a series of important documents to make the overall schedule and define the specific environmental protection responsibilities for the treatment of 56 lakes and reservoirs in Chongqing main urban area.

   Before 2017, the treatment of 56 lakes and reservoirs in Chongqing main urban area was led by Chongqing Environmental Protection Bureau and was basically completed by the environmental protection departments in 2016. According to the *Notice of General Office of Chongqing Municipal Committee of the CPC and Chongqing Municipal People's Government on Issuing Provisions on Chongqing Environmental Protection Work Responsibilities (Trial)* (YWBF [2016] No. 49), Chongqing Municipal Commission of Urban-rural Development shall take the lead to improve the water environment functions of 104 water bodies from 2017.

2. Work objectives and deployment.

The lake and reservoir treatment is one of the important contents for the eco-environmental protection and water pollution control strategy. Back in 2014, the goals for the lake and reservoir treatment in Chongqing were put forward in the *Notice of Chongqing Municipal People's Government on Issuing Work Programs for Atmospheric Pollution Control and Lake and Reservoir Treatment in Urban Function Core Area and Expansion Area* (YF [2014] No. 49) issued by Chongqing Municipal People's Government:

> In 2014, the water quality of 20 lakes and reservoirs such as the Baosheng Lake, Minzhu Lake, Leijiaqiao Reservoir, Caiyun Lake, Meihuashan Pond and Hualong Lake shall be improved from that in 2013, and the water quality of 36 lakes and reservoirs such as the Gailanxi Reservoir, Huayan Reservoir, Tuanjiehu Reservoir, Yinglonghu Reservoir, Changdangzi Reservoir and Baiyun Reservoir shall be stable with a good trend.

In the *Notice of Chongqing Municipal People's Government on Issuing Implementation Plan for Fully Implementing the State Council's Water Pollution Control Action Plan* issued in 2015, the objectives were clearly stated:

> By the end of 2015, the pollution treatment of 56 lakes and reservoirs in Chongqing main urban area shall be basically completed. By 2017, the ecosystem functions of the 56 lakes and reservoirs shall be basically restored; the black and odorous water in the urban built-up area shall be basically eliminated; and the black and odorous river cross sections in 38 key tributaries with a basin area over 500 square meters shall be basically eliminated.

In the *Notice of General Office of Chongqing Municipal People's Government on Issuing Work Program for Treatment of Urban Black and Odorous Water in Chongqing* (YFBF [2017] No. 85) issued by Chongqing Municipal People's Government in 2017, the time nodes were specified:

> By the end of November 2017, the pollution source control, pollutant interception and endogenous pollution treatment of 48 discovered segments of urban black and odorous water shall be completed with initial success in the treatment work to eliminate the black and odorous water; and the ecosystem functions of the 56 lakes and reservoirs in Chongqing main urban area shall be basically restored.

After taking over the black and odorous water treatment work in 2017, Chongqing Municipal Commission of Urban-rural Development attached great importance to it and successively issued documents such as *Letter on Forwarding the Notice on Conducting Evaluation on Urban Black and Odorous Water Treatment Effects* (YJH [2017] No. 206), *Notice of Chongqing Municipal Commission of*

Urban-rural Development on Make Full Preparation for Evaluation on Implementation of Water Pollution Control Action Plan (YJ [2017] No. 616), *Notice of Chongqing Municipal Commission of Urban-rural Development on Completing Rectification Measures Based on Feedback from Central Environmental Protection Inspectorate and Canceling Solved Problems* (YJ [2017] No. 615), in which it was required to achieve initial success in the treatment of 48 black and odorous water bodies by the end of 2017 and realize the goal of long-term clear water in 2018, so as to truly promote the black and odorous water treatment and evaluation work.

In 2017, for the water quality improvement of lakes and reservoirs, Chongqing Municipal Commission of Urban-rural Development also issued the *Notice of Chongqing Municipal Commission of Urban-rural Development on Further Consolidating Long-term Effects of 56 Lakes and Reservoirs in Main Urban Area* to require districts and counties to conduct in-depth investigation on the current situations and pollution factors of lakes and reservoirs, make objective evaluations on the original lake and reservoir treatment schemes, implementation and treatment effects and prepare evaluation reports on the lake and reservoir treatment effects. As for lakes and reservoirs, discovered in the investigation and effect evaluation, with problems such as poor water quality (black and odorous, or inferior to Class V), unsound treatment measures and no long-term mechanism established, rectification work shall be carried out immediately that special rectification schemes shall be established and implemented to ensure that the lakes and reservoirs can stably meet standards for Class V water quality.

According to the *Implementation Plan for Tough Battle of Pollution Control in Liangjiang New Area (2018–2020)* released by the CPC Working Committee of Chongqing Liangjiang New Area and Chongqing Liangjiang New Area Management Committee in 2018, the river chief system should be fully promoted by implementing the river chief system across the area, establishing a responsibility system, defining responsibilities of river chiefs and improving the collaboration mechanisms between the upstream and downstream and among departments, with the primary missions of "protecting water resources, controlling shorelines, improving water environment, restoring water ecology and realizing water safety". An integrated database for Liangjiang New Area water environment quality should be established for the integration of contents on engineering plans, construction schedules, problem identification and water quality monitoring of the planning, construction, city management departments and state-owned companies. Besides, online monitoring equipment should be installed in key lakes, reservoirs and rivers for the water quality monitoring and early warning, so as to integrate the data with the "Smart Liangjiang" information platform. The reconstruction of stormwater and sewage pipe networks around lakes and reservoirs should be carried, and the investigation and punishment on river-related illegal activities should be intensified, such as on illegal sewage discharge, illegal occupation of water shorelines, illegal sand excavation, illegal docks and illegal housing construction. The ecological environment safety evaluation should be carried on the Yulin River, and ecological environment protection schemes should be developed and implemented in basins of the Yunlin River, Houhe River, Heshuitan River, Chaoyang Stream, Gaodong River, etc. The environmental supervision system reform requirements should be implemented, and the pollution control long-term mechanism should be established and improved to conduct the whole-process supervision with the most stringent system. The reuse of reclaimed water should be encouraged in industrial enterprises to promote the increase of industrial water recycling rate in key industries.

3. Treatment results.

According to the *Notice on Further Implementing Relevant Work for Realizing Long-term Urban Clear Water* (YJ [2018] No. 95) issued by Chongqing Municipal Commission of Urban-rural Development in March 2018, the black and odorous water was basically eliminated in the abovementioned 48 urban segments in Chongqing, with the public satisfaction all over 90%, and the water quality of the 56 lakes and reservoirs was improved in 2017. The document also made corresponding arrangements for consolidating the treatment effects and further implementing relevant work for realizing long-term urban clear water. Meanwhile, Chongqing also required all districts and counties to strengthen measures such as water circulating and supplement, ecological restoration and "sponge city" construction. From 2018, the requirements for "sponge city" construction indicators shall be fully implemented in all construction links of urban water treatment projects to strictly control the stormwater runoff pollution, so as to improve the urban water environment from the source.

In the same year, Liangping District proposed the *Implementation Plan for Urban Black and Odorous Water Treatment Work in Liangping District, Chongqing* and other relevant plans, mainly for implementing source control, pollutant interception, endogenous source control, ecological restoration, circulating supplement and effect evaluation, which promoted the urban black and odorous water treatment orderly and effectively.

4.3.1.4 "One strategy for one river" program

According to the *Opinions on Fully Implementing River Chief System* issued by the General Office of the CPC Central Committee and General Office of the State Council on December 11, 2016, it was pointed out that the river chief system should be fully implemented. Moreover, at the Fifth Session of the Twelfth National People's Congress on March 5, 2017, Mr. Li Keqiang, Premier of the State Council, put forward to fully implement the river chief system and improve the ecological protection compensation mechanism in the government work report. The implementation of the "one strategy for one river" program is an indispensable part for fully implementing the river chief system and strengthening the treatment and protection of rivers and lakes, which helps to know about the river and lake health status, scientifically diagnose prominent problems of rivers and lakes, determine the objectives and main tasks for the treatment and protection of rivers and lakes and propose treatment and protection measures according to local conditions for aspects such as protection of river and lake water resources, waterline protection, water pollution control, integrated water environment treatment, water ecological restoration, long-term management and protection, law enforcement supervision and comprehensive function improvement.

During the implementation of the "one strategy for one river" program, protection and treatment measures shall be established according to the protection and treatment objectives and main tasks and with overall consideration of factors such as the local economic and social development level, work difficulty level and investment scale. In order to identify the priorities, implement various specific tasks and clarify the responsibility subjects for the implementation of the "one strategy for one river" program, the following four points shall be achieved:

(1) Highlighting key points and giving overall consideration: give overall consideration of such demands for protection of water resources, protection of river and lake resources, water resource control, make reasonable arrangements for treatment measures and implementation steps and give priorities to arranging treatment measures with obvious benefits and low implementation difficulty.
(2) Defining authorities and clarifying responsibilities: assign the treatment tasks and measures to each department and responsible person, break the overall objectives and tasks by river segment and year, make detailed implementation plans and designate the responsibility subjects.
(3) Implementing in steps and focusing on benefits: give priorities to the implementation of treatment measures featured with obvious results, wide benefited range and obvious effects on improving river water quality and promote the implementation of non-point pollution treatment projects with wide distribution scope and large quantities in steps by year.
(4) Making plans and promoting coordinately: make annual implementation plans and fund investment plans, pay attention to the fund balance and project balance during the implementation of annual plans and make arrangements for reasonable proportions of various projects to promote the projects coordinately.

Prominent problems for the management and protection of lakes and rivers can be properly solved by making overall plans for the upstream and downstream and both banks and implementing the "one strategy for one river" program according to actual conditions of different regions and different lakes and rivers. The overall objective of "unobstructed rivers, clear water, clean banks and beautiful landscape" for rivers can be achieved through full implementation and deepening of the river chief system, integrated treatment, effective control and treatment of water pollution, comprehensive treatment of water environment, establishment of long-term management and protection mechanism and intensified law enforcement supervision.

In 2018, Nan'an District of Chongqing completed the preparation of the *Implementation Plan for Special Rectification of Stormwater and Sewage Drainage Pipeline Construction in Nan'an District during the "13th Five-year Plan" Period* (2016–2020) and the construction of main works and supporting trunk networks of Donggang sewage treatment plant, steadily promoted the Class I(A) standard upgrading and reconstruction project of Jiguanshi sewage treatment plant, actively carried out the survey of stormwater and sewage pipelines, overfulfilled the stormwater and sewage pipeline construction plan released by Chongqing Municipal Commission of Urban-rural Development in 2017 and completed the special planning for construction. Nan'an District also established the organization structure of the river chief system, established the river chief system management information system, organized the preparation and implementation of the "one strategy for one river" program, established "one file for one river" and put funds in place for the management and protection of lakes and rivers in each sub-district and town. At present, the patrol information of each river segment is reported on time through the information system; the "one strategy for one river (reservoir)" programs for two rivers and six reservoirs at the district level have been prepared; and the special fund with an amount over 700,000 yuan has been allocated for the cleaning of flotsam on water surface and waste on river (reservoir) banks.

4.3.2 Improvement of management mechanism

4.3.2.1 Government management

1. Improve integrated water environment treatment system and mechanism.
   The integrated water environment treatment is a kind of systematic and comprehensive work involving multiple fields, departments and links, so a systematic, complete, scientific, standard and effective responsibility system is needed. First, the treatment and protection responsibilities for each water body shall be assigned to a specific responsible person or responsible institution by giving full play to the long-term management function of the "river chief system". Second, the responsibility system and reward and punishment system shall be combined to intensify the reward and punishment based on evaluation, strengthen the supervision and management on "river chiefs" to promote better performance of "river chiefs" and motivate the subject consciousness and work initiatives of "river chiefs". Third, the supporting mechanism shall be established and improved. As the water treatment and protection involves multiple departments and multiple perspectives, smooth systems and mechanisms shall be established by connecting the "river chief system" with other systems and improving supporting systems such as joint conference, information exchange and sharing, inspection and supervision systems, so as to effectively solve key and difficult problems for the protection of lakes and rivers. It is hard to effectively promote the water environment protection work only relying on the efforts of environmental protection departments, so the water environment pollution control responsibilities of environmental protection, land resources, water, marine and fishery departments shall be integrated to establish a multi-departmental collaboration mechanism to realize unified planning, standard, environmental assessment, monitoring and law enforcement.
2. Strengthen overall planning for integrated water environment treatment.
   Attention shall be paid to the water system connectivity to avoid "focusing on local areas and ignoring the overall situation". Like the old saying: "Running water never becomes putrid, and a rolling stone gathers no moss". The liquidity fragmentation of rivers, lakes and wetlands will lead to increased difficulty in water pollution treatment and decreased water ecological functions and even loss of their value. Attention shall also be paid to the overall planning for the construction of sewage and waste treatment facilities to avoid "focusing on construction and ignoring planning". The key to water pollution control is to control pollution sources. The overall planning for underground pipe networks shall be made, and the scientific and reasonable layout for the construction of sewage treatment, domestic waste treatment and construction waste recycling treatment facilities shall be designed with the combination of shanty towns reconstruction, urban village reconstruction and road reconstruction in old urban areas.

3. Improve accountability mechanism for government responsibility for black and odorous water treatment.

Accountability facilitates the performance of duties. The accountability mechanism is a necessary means to ensure that government officials perform their duties in the black and odorous water treatment. Regulations and laws on the investigation of responsibility for eco-environmental damage shall be established and improved; the responsibility determination mechanism shall be strictly implemented; eco-environmental responsibility lists shall be established and the accountability standards, subjects and procedures shall be specified to provide bases for the accountability. A strict accountability mechanism shall be implemented for the eco-environmental protection and the lifetime responsibility system shall be implemented for the eco-environmental damage. A cohesive mechanism between administrative law enforcement and criminal justice shall be improved. The accountability ways shall be improved by expanding variant accountability system, reinforcing accountability by the National People's Congress, strictly implementing judicial accountability and encouraging media accountability.

4.3.2.2 Establishment of laws on black and odorous water treatment

1. Special legislation for black and odorous water treatment.

The special legislation for urban black and odorous water treatment shall be carried out. First, "one law for one city" shall be implemented. As mentioned above, the distribution and treatment of black and odorous water have distinct geographic and regional characteristics, so the legislation scope shall be narrowed as far as possible to make the legislation suitable for the local black and odorous water treatment to the greatest extent. In accordance with relevant provisions of the *Legislation Law*, all "cities at the prefecture level or above" shall have the legislative authority in accordance with the law. Therefore, the legislation for black and odorous water treatment shall be positioned as Level 1 law of cities at the prefecture level or above and the "one law for one city" shall be implemented. In such way, the legislation effectiveness can be guaranteed and the legislation can be enacted according to local conditions to ensure its scientificity and practicality. Next, the special legislation shall be realized. From the above analysis, it is obvious that current legislation on water pollution and policy documents specific to black and odorous water can't meet the actual needs of black and odorous water treatment nationwide. The legislative mode for urban black and odorous water treatment shall be clearly defined as special legislation on black and odorous water treatment, so as to guarantee the water treatment objective of "long-term clear water" through legislation, scientifically establish the legal system for the black and odorous water treatment and fully implement the concept of systematic water treatment based on the systematic legislation and improve government performance of duties, urge enterprises to act legally and ensure the public engagement through the legislative authority. Therefore, a special legislative mode shall be adopted, in which special legislation on treatment of urban black and odorous water in different places shall be established according to corresponding actual conditions.

2. Legal principles for black and odorous water treatment.

Legal principles refer to the fundamental truths and principles of laws or comprehensive principles or starting points that provide basis or origin for other legal elements. As the "tools" to understand legislative purposes, guide the enforcement of legal systems, apply bases for trial and correct injustices, the legal principles play a particularly important role in a law. In combination with current water laws in China and abroad, the legislation on urban black and odorous water shall adhere to the following three principles:

① Principle of government domination.

As the systematic engineering of local area, the black and odorous water treatment shall be subject to unified organization, leadership and decision-making of the people's governments at all levels. Therefore, the conventional thinking of environmental protection legislation that focuses on "enterprise control" and "pollution control" shall be changed in the legislation on urban black and odorous water treatment. On the one hand, the authority for organization, leadership and coordination of local people's government shall be emphasized. On the other hand, the supervision on and accountability for local people's government's performance of duties shall be stressed, so as to urge people's governments at all levels to play their due roles in the black and odorous water treatment through the two measures at the same time.

② Principle of department collaboration.

The black and odorous water treatment is not limited to the water pollution control, so the urban black and odorous water treatment is not merely the duty of one government department but involves multiple authority (duties) of relevant government departments. Therefore, the legislation on urban black and odorous water treatment shall define the duties of responsible departments involved in the treatment work and specify the division of responsibilities. Meanwhile, the collaboration among government departments shall be enhanced to form the departmental collaboration, so as to achieve the scientific treatment situation that government departments perform their duties and cooperate with each other to make concerted efforts in the black and odorous water treatment.

③ Principle of systematic treatment.

In order to solve the black and odorous water problem, the combined force of legal systems for "water control" and "bank protection", "discharge reduction" and "capacity increase" shall be formed, which necessarily requires a principle of systematic treatment established in the legislation to implement and express the above concept and measures in the form of laws and norms. Meanwhile, the systematic treatment also needs to manage the relationship between water and other elements in the ecosystem. Therefore, the idea of overall situation shall be established in the legislation to give an overall consideration of the control of mountains, forests, farmland, lakes and grasses. Moreover, on the basis of sticking to the basic principle of systematic treatment, overall planning for mountain, water, forest, farmland, lake and grass shall be made through the design of a series of legal systems for pollution source control, pollution discharge behaviors, waterline management and maintenance, water treatment, ecological protection and restoration, so as to realize comprehensive treatment of black and odorous water finally.

4.3.3 Increase public engagement

The public is the biggest "victim" of black and odorous water. Therefore, the speech right and judging right shall be given to the public during the black and odorous water treatment, and the public satisfaction with treatment results shall be the primary criterion for judgment of the black and odorous water treatment results. Besides, it shall be guaranteed that the public can participate in the whole process of the water treatment. Therefore, the public engagement's important role in the long-term maintenance of black and odorous water treatment effects shall be highly valued in the legislation on urban black and odorous water treatment to effectively mobilize the public engagement, expand the information disclosure channels for urban black and odorous water treatment, improve the government environment information way, encourage the public engagement and accept the public review and public supervision, so as to ensure the public can timely know about relevant water treatment information and have channels to participate in the treatment, forming the new situation of multi-party joint treatment of black and odorous water.

The force of government environmental protection departments is far from enough to enhance the black and odorous water treatment and environmental supervision, ensure effective treatment of environmental pollution and ecological damage and implement the system of government's primary responsibility for ecology only relying on the force of government environmental protection departments. Public participation is an inevitable demand of environmental protection under the democratic system. The social participation in ecological treatment is highly valued all over the world, and China also encourages the public to participate in the supervision on ecological treatment and has clearly stipulated in the *Law on Environmental Impact Assessment* (2003) that "the state shall encourage relevant entities, experts and the public to participate in environmental impact assessment by proper means" and also has specified concrete participation ways. Therefore, relevant systems and measures for the public engagement shall be improved for the purpose of black and odorous water treatment.

4.3.3.1 Establish and improve public engagement ways and government response mechanism for black and odorous water treatment

As for the public engagement ways, first, the hearing system for the black and odorous water treatment and environmental impact assessment shall be improved. Where any major decision about black and odorous water treatment may affect the ecological environment, the government shall hold a hearing to let the public express their personal opinions on ecological protection and assessment to maintain environmental rights of the public; second, the information disclosure system for black and odorous water treatment shall be established to timely and correctly disclose information about black and odorous water treatment and environmental information monitored at monitoring stations to satisfy the public right to know, especially the information about major pollution emergencies shall be timely disclosed to the public; third, NGOs and voluntary activities for black and odorous water treatment shall be encouraged and the NGOs' comments and suggestions on black and odorous water treatment shall be carefully listened to, so as to unite with NGOs to build a social ecological "firewall", etc. As for the government response mechanism, on the one hand, the public supervision procedures shall be specified, including the contents of public supervision on ecological environment, departments receiving the public reporting, concrete ways of reporting, departments responsible for disposal or interpretation and explanation to the public, etc.; on the other hand, the effectiveness of public supervision shall be ensured. Governments shall timely respond the public supervision, immediately stop illegal acts of enterprises or individuals, handle in accordance with the law and publicize the handling results, even give rewards for the public supervision and clearly express governments' attention to and support for the public supervision to stimulate the enthusiasm for and effectiveness of public supervision.

4.3.3.2 Establish and improve public reporting system to promote the public supervision on black and odorous water treatment

One of the motivations for the performance of ecological protection legal systems and environments' ecological responsibility is from supervision of the public, especially residents affected by environmental pollution will definitely report to corresponding departments or expose to the media, so as to supervise and urge governments to stop pollution behaviors of enterprises. Therefore, the public engagement and establishment of reporting system will become the true motivations and pressure for ecological protection of governments and enterprises.

4.3.3.3 Enhance the public awareness of ecological responsibility and implement the veto power for governments' responsibility for ecology

The enhanced environmental awareness and broad participation of the public are important foundations for environmental protection. Generally speaking, the environmental awareness mainly refers to the people's consciousness of participating in environmental protection according to their basic ecological values on the basis of their cognition of environment conditions and environmental regulations, reflected in practical actions for environmental protection. So far, the public has low environmental awareness in China and has certain bias in environmental cognition of cognitive or inadequate cognition of governments' environmental behaviors, resulting in their inadequate attention to environmental protection and ecological treatment, let alone the supervision on governments' responsibility for ecological treatment and exercise of ecological veto power. Therefore, governments shall actively conduct the ecological protection publicity and education to local residents to stimulate and enhance their awareness of and participation in ecological environmental protection and intensify the public awareness of ecological rights, so as to realize strong supervision on governments' responsibility for ecology. Especially, the publicity of environmental awareness shall be conducted in key areas around lakes and in key enterprises to let them realize that the incompetent environment treatment of governments is a kind of default and the governments shall be held accountable accordingly, thus the ecological treatment responsibility of governments can be truly made public and subject to public supervision to accept the public inspection and questioning.

Chapter 5

Practice of source treatment

5.1 Planning and implementation of sponge city

5.1.1 Significance of sponge city

Urbanization is a powerful engine for maintaining sustainable and sound economic development, a strong support for promoting coordinated regional development and a necessary demand for promoting all-round social progress. However, during the rapid urbanization, urban development also faces tremendous environment and resource pressure. Under such circumstance, the extensive growth-type urban development pattern is hard to sustain, so China's urbanization must enter a new stage of transformation development centered on quality improvement. For that purpose, we must adhere to the new-type urbanization development path and coordinate the contradiction between urbanization and protection of environment and resources, so as to realize the sustainable development. The construction of sponge cities with natural accumulation, natural infiltration and natural purification functions is an important part of the ecological civilization construction, an important manifestation of the co-ordinated development of urbanization and environmental resources and also a major task for China's urban construction in future.

As the name suggests, sponge city refers to a city that could be functioned as a sponge that has great "resilience" in adapting to environmental changes and coping with natural disasters, such as to absorb water, store, infiltrate and purify water when it rains and to "release" and utilize the stored water as needed. In the sponge city construction, the principle of ecology priority shall be followed to combine natural approaches with artificial measures, and on the premise of ensuring the safety of urban drainage and waterlogging control, the retention, storage, infiltration and purification of stormwater in the urban area shall be realized to the greatest extent, so as to promote the stormwater resource utilization and eco-environmental protection. During the sponge city construction, the natural precipitation, surface water and groundwater systems shall be taken into overall consideration, the water supply, drainage and other water recycling links shall be coordinated, and the complexity and long-term performance shall also be taken into account.

A sponge city is mainly constructed by the following ways. The first one is protection of the city's original ecosystem by protecting the original rivers, lakes, wetlands, ponds, ditches, channels and other water ecologically sensitive areas to the maximum extent, conserving enough forests, grassland, lakes and wetlands for water conservation and coping with heavy rainfall and maintaining the city's natural hydrological characteristics before development, which are basic requirements for the sponge city construction; the second one is ecological rehabilitation and restoration by rehabilitating and restoring water bodies and other natural circumstance destructed under the traditional extensive urban construction pattern through ecological means and maintaining a certain proportion of ecological space; the third is low-impact development by reasonably controlling the development intensity, reserving enough ecological land in the city, controlling the proportion of impermeable area in the city and minimizing destruction of the original water ecological environment according to the concept of minimizing the development and construction's impact on urban ecological environment and also by appropriate excavation of rivers, lakes, ditches and channels and increase of water area to promote the stormwater storage, infiltration and purification.

The "sponge city" construction can effectively utilize abundant stormwater resources while preventing and controlling flood. The "sponge city" construction is aimed at improving urban drainage and optimizing urban construction to make cities better serve the society. Therefore, the most important and valuable meaning of "sponge city" construction is to effectively utilize abundant stormwater resources while preventing and controlling flood. From the perspective of flood control, the "sponge city" construction is to relieve the urban drainage pressure and avoid the urban flood control system not meeting the use requirements in case of heavy rainfall, which are the primary purposes and fundamental objectives of the "sponge city" construction. From the perspective of stormwater resource utilization, the greatest value of "sponge city" construction is to help cities effectively utilize stormwater resources, and that's where the name "sponge city" comes from. Just as the name implies, the "sponge" not only can absorb water but also drain water, which absorbs water to relieve urban drainage pressure during flood period and drains water to supply water for life and production.

5.1.2 Planning and implementation of sponge measures

The sponge city planning is mainly to solve problems and define objectives. Therefore, the planning for measures may be detailed from the aspects of water ecology, water safety, water resources and water environment, and then summarized and optimized.

5.1.2.1 Planning for water resource utilization system

Combined with the urban water resource distribution and water supply engineering, centered on urban water resource objectives, the water sources shall be strictly protected and technical schemes and implementation approaches for comprehensive utilization of reclaimed water and stormwater resources shall be prepared to improve the exploration and utilization level of local water resources, enhance the water supply safety and specify the layouts, land, functions and scale of major municipal facilities such as water source conservation areas, reclaimed water plants, small reservoirs, ponds and stormwater comprehensive utilization facilities that may occupy land independently shall be specified.

5.1.2.2 Planning for integrated water environment treatment

Measures for water environment treatment shall be put forward according to local conditions based on comprehensive analysis and evaluation of current situations of urban water environment and overall consideration of multiple relationships between short term and long term, ecology and safety, landscape and functions, etc. In combination with current situations, capacity and function zoning of urban water environment and centered on the control objective for total urban water environment capacity, the approaches for meeting standards and implementation approaches for various technical facilities shall be specified, and the water environment treatment system technical schemes shall be prepared, including schemes for point source pollution monitoring and control, non-point source pollution control (source, intermediate and end) and improvement of water self-purification capacity. Measures such as various types of water diversion for diluting pollution and connection of rivers and lakes shall be resolutely banned on account of hydrodynamic force restoration.

On the basis of full analysis and demonstration, the combined systems require reconstruction recently shall be identified by sorting existing urban sewer systems. As for those meeting conditions for reconstruction into separate rain and sewage systems, the transformation shall be intensified. As for those not meeting the reconstruction conditions, pollutant interception measures shall be taken properly and the annual average number of sewage overflow times and total sewage overflow volume of combined systems shall be controlled comprehensively in combination with the construction of sponge cities, regulation and storage facilities.

The layouts, land, functions and scale of major facilities such as sewage treatment plants, sewage (pollutant interception) regulation facilities and wetlands that may occupy land independently shall be specified and optimized, with overall consideration of supplement of ecological water with reclaimed water from sewage treatment plants, restoration of hydrodynamic force and review of water environment goal attainability.

5.1.2.3 Planning for water ecological restoration

Combined with urban runoff generation and convergence characteristics and current water system situation and centered on urban water ecology goals, the approaches for meeting standards shall be specified, the breakdown schemes for annual total runoff control rate and layout schemes for ecological shoreline restoration and protection shall be prepared and the functions, forms and overall control requirements for bank lines of key water systems shall be specified.

The protection and restoration of natural forms of ponds, rivers, lakes, wetlands and other water bodies in cities shall be intensified; the water ecological environment once subject to destruction for channel straightening and hardening shall be identified and analyzed; as for those meeting reconstruction conditions, the technical measures and schedules for ecological restoration shall be put forward to reconstruct channelized rivers, rebuild healthy and natural curved river bank lines, restore natural pools and shoals and implement ecological restoration. The rebuilding of natural bank lines can restore the hydrodynamic force and biodiversity to make use of the natural purification and restoration functions of rivers.

5.1.2.4 Planning for water safety guarantee

The current water environment situation shall be fully analyzed to evaluate the city's drainage capacity and water environment risk. Combined with the treatment of urban areas vulnerable to waterlogging, current situation of and planning for drainage and waterlogging prevention engineering, centered on the goal for urban water safety, technical schemes for urban drainage and waterlogging prevention system shall be developed with overall consideration of the combination of multiple measures such as infiltration, retention, storage, purification, utilization and drainage; and the runoff control goals, implementation approaches, standards and construction requirements of the source runoff control system, pipe system and waterlogging control system shall be defined, respectively.

As for existing built-up areas, drainage zones shall be reasonably optimized with the treatment of places vulnerable to waterlogging as the breakthrough point to gradually reconstruct urban drainage trunk systems, raise construction standards and systematically improve the urban drainage and waterlogging prevention capacity.

The layouts, land, functions and scale of major municipal facilities such as regulating storage ponds, flood retention areas and pump stations that may occupy land independently shall be specified. Control requirements for properties of vertical areas and areas vulnerable to waterlogging shall also be specified.

5.2 Special planning for sponge city in Chongqing main urban area

According to the special planning for sponge city in Chongqing main urban area, the planned area is 5,473 km^2, including construction land area of 1,188 km^2, with the planned population about 12 million. The planning period is consistent with the overall urban and rural planning of Chongqing, with the short-term planning period till 2020. The planning shall focus on the cluster where the urban construction land is located at, with an area of 1,712 km^2, including construction land area of 1,158 km^2 and regional infrastructure area of 38 km^2.

According to natural characteristics and environmental conditions in the main urban area, measures such as "infiltration, retention, storage, purification, utilization and drainage" shall be comprehensively adopted to absorb and utilize 75% rainfall on the spot, improve the ecological pattern, improve water environment, restore water ecology, strengthen water safety and guarantee water resources to build a "3D sponge city with mountainous characteristics" and achieve the goals of "no black and odorous water, no puddle in case of light rain, no waterlogging in case heavy rain and alleviative heat island effect". More than 20% of the urban built-up area shall achieve the goals by 2020 and more than 80% of the urban built-up area shall achieve the goals by 2030.

Sticking to the principle of adaptation to local conditions, the old urban area shall be problem-oriented and focus on solving problems such as runoff pollution, black and odorous water, local puddles and large-area hardening; the new urban area shall be goal-oriented and give priority to protection of natural ecology, comprehensive balance of the natural ecological protection, urban development and economic input, as well as improvement of comprehensive benefits of sponge city construction. Overall planning for natural ecological functions and manual intervention functions shall be made that focuses on the source reduction and combines with process control and end treatment to form a complete stormwater integrated management system.

5.2.1 Analysis on problems and demands

5.2.1.1 Water ecology problems

1. Problem of water and soil loss.

   Chongqing section of the Three Gorges Reservoir Region is divided into southwest earth-rock mountainous area among the water and soil loss types in China and is a national key supervision area for water and soil loss and key treatment area for water and soil loss, accounting for 80% of total area of the Three Gorges Reservoir Region and covering most of the Reservoir Region, so the solving of the water and soil loss problem is of special strategic significance to the long-term safe operation of the Three Gorges Key Water Conservancy Project and the flood control and ecological safety in the lower reaches of the Yangtze River.

   According to the remote sensing survey in 2005, the water and soil loss area in Chongqing was 40,000 km^2, accounting for 48.55% of the total area of Chongqing, with average soil erosion modulus of 3,641.95 t/(km^2·a) and total soil erosion amount of 146 million t/a. Among them, the water and soil loss area in the Three Gorges Reservoir Region was 23,800 km^2, accounting for 51.71% of the total area, with average soil erosion modulus of 3,738.51 t/(km^2/a) and total soil erosion amount of 89.24 million t/a. Chongqing invests a lot of money in the water and soil loss treatment every year. According to the data released in the *Bulletin for Water and Soil Conservation Conditions in the First National Census for Water* in May 2013, the water erosion area in Chongqing was 31,363 km^2, accounting for 38.07% of the total area (Table 5.1).

   From Figure 5.1, water and soil loss insensitive areas and low sensitive account for the majority of Chongqing main urban area, with moderately sensitive and relatively sensitive areas mainly distributed to the north of Jialing River and Yangtze River and highly sensitive areas mainly distributed in Beibei District, Yubei District, Nan'an District and small part of Banan District.

Table 5.1 Annual water and soil loss treatment area during 2004–2013

Year	Annual water and soil loss treatment area (km^2)
2004	2,201.0
2005	2,514.21
2006	2,533
2007	2,612.67
2008	2,538.7
2009	2,815.2
2010	2,418.73
2011	3,175
2012	1,675
2013	1,527

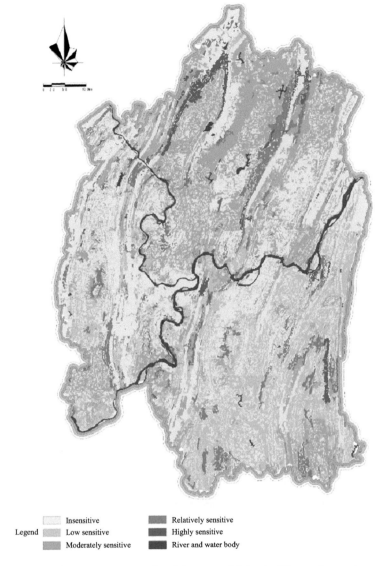

Legend

Insensitive   Relatively sensitive
Low sensitive   Highly sensitive
Moderately sensitive   River and water body

Figure 5.1 Analysis on water and soil loss sensitivity in Chongqing main urban area.

The water and soil loss will lead to decline in soil fertility and loss of large amount of fertile topsoil, reservoir deposition, riverbed elevation, navigation capacity reduction and flood disaster, threat to the safety of industrial, mining and traffic facilities as well as ecological environment deterioration.

2. Environmental problems in ecologically sensitive areas.

Ecologically sensitive areas refer to areas with special sensitivity to human production and life activities or with potential natural disaster impact, and areas especially susceptible to improper human development activities, resulting in negative ecological effects. Ecologically sensitive areas include all elements belonging to ecology, such as creatures, ecological environment, water resources, atmosphere, soil, geology, landforms and environmental pollution. As the areas with the most intense ecological environment changes and most prone to ecological problems in a region, ecologically sensitive areas are also the key areas for sustainable development of the regional ecosystem and integrated treatment of ecological environment.

As a typical city with mountains and rivers, Chongqing has developed water systems and typical landforms. According to the distinctive environment in Chongqing, the ecologically sensitive area types mainly include karst ecologically sensitive area and typical subtropical karst water source conservation area.

The karst environment is a unique ecosystem with a series of fragile ecology characteristics, such as low environmental capacity, small biomass, high sensitivity to ecological environment system variation, weak anti-interference capacity and poor stability. Chongqing is located at the edge of the second terrain level of China, with well-developed karst landforms. The karst sensitive areas in Chongqing main urban area are mainly distributed in the Jinyun Mountains, Zhongliang Mountains, Tongluo Mountains, Mingyue Mountains, Dongwenquan Mountains, etc.

Water source conservation areas refer to those delineated by the country for special protection of some very important water bodies and are key areas for pollution control. The water source conservation areas in Chongqing are mainly distributed in tributary areas (Yubei District, Beibei District) in the upper reaches of Jialing River and some tributary areas (Banan District) in the lower reaches of the Yangtze River.

5.2.1.2 Water environment problems

At present, Chongqing water environment problems mainly involve point source pollution and non-point source pollution problems. The point source pollution means that pollutants are discharged into water from centralized places, mainly including domestic pollution and industrial pollution. As for Chongqing main urban area, the point source pollution mainly has the following characteristics:

① There are relatively less industrial enterprises in Chongqing main urban area and the management on sewage discharge of industrial enterprises is relatively stricter that the industrial wastewater can only be discharged into the urban sewage pipe network after being treated and meeting relevant discharge standards. Therefore, the point source pollution in Chongqing main urban area is mainly domestic pollution.
② The sewage network coverage rate has not reached 100% yet, so the sewage in some areas is still discharged in a unorganized way or discharged into storm sewer.
③ Some old sewage pipes exist the problem of serious leakage.
④ The treatment capacity of existing sewage treatment plants is limited.

Therefore, part of the domestic sewage is discharged into water bodies due to the inadequate sewage collection and treatment facilities, resulting in certain pollution to the surrounding water environment.

With the rapid progress of urbanization, the urban runoff pollution caused by storm runoff scouring has become an important threat to the water quality safety of receiving water. Due to the strong randomness and many influence factors of urban runoff pollution, the research results in different regions are rarely consistent. Due to the rugged and varied landforms in a mountainous city, the occurrence rule of runoff pollution is more complex.

According to the rainfall runoff monitoring results for different underlying surfaces and basins in Chongqing, the event mean concentrations (EMCs) of total suspended solids (TSS) and chemical oxygen demand (COD) of urban trunk roads are significantly higher than those of roads in living quarters, commercial districts, concrete roofs, tiled roofs and campus integrated collection areas. Meanwhile, the EMCs of total nitrogen (TN) of commercial districts and urban trunk roads are close to 7.1–8.9 mg/L, which are higher than those of concrete roofs, tiled roofs and campus integrated collection areas. The pollution loads of TSS, COD and total phosphorus (TP) of urban trunk roads are 589, 404 and 1.0 t/(km^2/y), respectively, and are major contributors to the TSS, COD and TP urban runoff pollution loads; meanwhile, the TN and NH$_3$-N urban runoff pollution loads are mainly from urban trunk roads, commercial districts and roads in living quarters and are 6.6–8.5 and 4.1–4.5 t/(km^2/y), respectively.

5.2.1.3 Water resource problems

Chongqing has abundant transit water resources, but less utilizable water resources, with the multi-year average precipitation of about 1,184 mm, surface water resources of 56.7 billion m^3, groundwater resources of 9.6 billion m^3, average water production coefficient of 0.58 and average water production modulus of 730,000 m^3/km^2. The local water resource per capita is only 1,644 m^3. According to international "water stress indexes": areas with the water resources per capita of 1,000–1,700 m^3 are in a state of moderate water shortage, such as Chongqing. Especially, the precipitation in Chongqing is easily affected by the climate. In case of moderate droughts, among the total water resources in Chongqing, the water resources available for direct use are about 12.1 billion m^3, with the water resource per capita about 393 m^3.

The spatial and temporal distribution of water resources in Chongqing is very uneven, with 70% water in flood season and 30% in non-flood season, most water in the east and less in the west; the hills and flatlands have large populations but less water and small areas, while high mountains and high hills have small populations, less water but large areas; the soil moisture layer is thin and the water retaining capacity is poor.

Chongqing has a large terrain elevation difference, with the main urban area at the height above sea level of 200–550 m. Therefore, it costs a lot to lift water from Jialing River and Yangtze River. Meanwhile, the utilization of water resources there is characterized by high difficulty and high cost due to deeply incised valleys, and a large amount of stormwater is rapidly drained via hardened road surfaces, resulting in a waste of resources. Due to the poor water retaining capacity of mountainous city's original landforms after rain and large water demand for landscaping irrigation and road and square washing, the municipal water supply via river water lifting in high-elevation areas accounts for about 10% of total water consumption, resulting in high costs. In Chongqing, the surface water utilization accounts for 7.8% of local total water resources, and the groundwater utilization accounts for 1.5% of local total groundwater resources. Therefore, Chongqing is highly dependent on water conservancy projects for the water resource utilization, belonging to the engineering-caused water shortage area.

5.2.1.4 Water safety problems

Chongqing has a multi-year average precipitation of 1,208 mm, with the precipitation mainly concentrated in flood season (May to September), accounting for 69% of annual total precipitation. The precipitation is distributed very unevenly with an early rainfall peak. Chongqing main urban area has a serious situation of waterlogging, with 219 places vulnerable to waterlogging.

Chongqing is a typical mountainous city, with widely distributed mountains and hills, rugged terrain and large elevation difference. The precipitation is greatly influenced by the terrains. Besides, the unique terrains and landforms of Chongqing also make the water have large potential energy difference, strong liquidity and gather from high elevation to low elevation, forming an asymmetrical centripetal water system network with many tributaries. In case of large-area precipitation or local heavy precipitation, the stormwater will rapidly gather in low-lying areas, which are vulnerable to waterlogging due to small gradient of drainage lines, resulting in serious flood disasters.

In recent years, with the continuously accelerated urbanization process in Chongqing main urban area and constantly expanded urban size and under the background of climate change and rapid development of urbanization, the intensity and distribution characteristics of regional short-duration heavy precipitation have been changed significantly and the intensity of extreme precipitation events has increased, such as the heavy precipitation events on July 17, 2007, August 4, 2009, June 9, 2013 and June 24, 2016, resulting in poor drainage, waterlogging and traffic jams and causing serious social impacts and economic losses.

5.2.1.5 Analysis on construction space guarantee of sponge city

The construction land in a mountainous city is narrow, and mountain cutting is always needed to expand the land. The land exploitation and utilization have more direct impacts on the natural environment of mountainous region, and the original sponge structure of water infiltration, water retention and water holding is more vulnerable to damage. As for the construction space guarantee, the plane extension and 3D structure of the city sponge structures shall be taken into full consideration. The development areas shall be strictly defined or controlled in the plane to strictly control the urban development and construction boundaries, and in the vertical direction, the mountainous terrains shall be combined with to construct the sponge structure in mountainous region, optimize water system corridors and strengthen the construction of mountainous region sponges with the functions of "water infiltration, retention, storage, purification, utilization and drainage".

The main urban area has a good ecological basis for the construction of sponge areas and corridors. The urban green space system has a strong natural ecological mechanism and good overall ecological effects. The urban land use is complex, with large proportions of irregular green spaces and lots of green wedges and belts. However, due to the land shortage, the urban green spaces are not great in number, which are separated and randomly and unevenly distributed with large elevation differences. Due to the rugged, uneven, winding and varied mountainous terrain, the green spaces in a mountainous city are also layered and scattered randomly, naturally forming a 3D green space system and conducive to the stability and benign cycle of the urban ecological sponge system.

Water systems are the basic carriers and flowing veins connecting "sponges in mountainous region", and the natural drainage system formed by urban water systems is an important part of the ecological stormwater management of a sponge city. The main urban area has many hills, low mountains and mountainous landforms, and complex and varied terrains, with crisscross water parting lines at different levels. Water systems and their waterfront spaces consist the typical linear sponge corridors of the mountainous city, which are flowing carriers maintaining the water circulation of mountainous region sponge bodies and key links connecting mountains, green spaces, urban spaces and water ecological sponges.

5.2.1.6 Analysis on demands of water problems

According to the planning strategy for the beautiful city with mountains and rivers of Chongqing, the ecological environment protection and construction of shall be strengthened, the pollution prevention and treatment shall be intensified and the low-carbon lifestyle and green development shall be advocated. The ecological civilization construction shall be carried out in the whole process of urban-rural planning and construction management, economic and social development and lives of residents. The construction of beautiful city with mountains and rivers is an important carrier and component of the ecological civilization construction.

As a typical city with mountains and rivers, Chongqing has developed water systems and wide urban green space area. However, due to the population growth and rapid development since being a municipality directly under the central government, Chongqing has encountered many problems such as ecological damage, water environment pollution, water shortage and urban waterlogging during the urban development. In order to solve the contradiction between development and resources and environment and achieve the goal for urban development, Chongqing shall take the road of intensive and low-carbon ecological development. Therefore, the construction of low impact development (LID) stormwater system according to the sponge city construction requirements is the only way to solve the problems of water ecology, water safety, water environment and water resources in Chongqing main urban area.

1. Further improve and upgrade water eco-environmental quality, restore natural water ecosystem and reduce water and soil loss by the sponge city construction. According to the concept and requirements of the sponge city construction, the original rivers, lakes, wetlands and other water ecologically sensitive areas shall be protected to the maximum extent to maintain the natural hydrological characteristics before urban development. Meanwhile, the proportion of impermeable area in the city shall be controlled to minimize the original water ecological environment destruction caused by urban development and construction. Besides,

the water bodies and other natural circumstance destructed under the traditional extensive urban construction pattern shall be rehabilitated and restored through ecological means.

2. Further intensify the urban flood control and drainage system by the sponge city construction. The urban waterlogging shall be avoided by constructing regulating storage ponds or increasing regulating storage water bodies as needed, discharging water by opening sluice gates at low water level before stormwater to make room for regulation and storage, and storing water by closing sluice gates at high water level. Meanwhile, the stormwater storage, infiltration and purification shall be promoted to improve the service capacity of urban storm sewer system and excessive storm runoff discharge system to a certain degree and give full play to the water regulation and storage functions of natural ecosystem, so as to effectively alleviate the urban flood drainage pressure.

3. Further alleviate the water environment treatment pressure, improve water environment and eliminate black and odorous water by the sponge city construction. The LID measures such as green roof, eco-grass ditch and stormwater garden shall be adopted to intercept non-point source pollution while storing and retaining stormwater. Constructed wetlands shall be built as needed in combination with original wetlands, and the purification function of natural ecosystem shall be fully used to cut the pollutants enter rivers to an environmental allowable range, so as to alleviate the problem of serious urban water pollution.

4. Further guarantee the water resource safety by the sponge city construction. In urban construction areas, the lakes, ponds, reservoirs, pools and others shall be fully used to retain, store and utilize stormwater, and stormwater and reclaimed water shall be utilized for industrial, agricultural and ecological uses as much as possible, while the quality surface water shall be utilized for domestic use, which can reduce the urban flood risk and also alleviate the realistic problem of water resource shortage.

5.2.2 Sponge space pattern building and function zoning

5.2.2.1 Mountains, rivers, forests, farmland and lakes

The mountain area in the main urban area totals about $1,634 \text{km}^2$, accounting for about 30% of the planned area. The mountain geographical units in Chongqing main urban area are divided into middle mountains, low mountains and hills and include paralleled mountains, isolated high mountains, mountains in the main urban area and scarps from the perspective of terrain conditions. Middle and low mountains are important ecological protective screens and "green lungs" of the main urban area, playing important roles in conserving water and soil, purifying air, regulating climate, resisting natural hazards, mitigating urban heat island effects, etc. Mountains in the main urban area mainly include four mountains, namely, the Jinyun Mountain, Zhongliang Mountain (including Longwangdong Mountain), Tongluo Mountain and Mingyue Mountain; two ridges, namely, the central ridge line of the Pipa Mountain–Eling Mountain–Pingding Mountain, and north ridge line of the Longwangdong Mountain–Zhaomu Mountain–Shizi Mountain and its offsets; and 40 important mountains in the main urban area, such as the Qiaoping Mountain, Yunzhuan Mountain, Zhaishanping Mountain and Yuntai Mountain.

The water area in the main urban area totals about 198 km², accounting for about 3.6% of the planned area. Chongqing main urban area is crisscrossed by many rivers and dense with water networks, and all the rivers belong to the Yangtze River system. The Yangtze River and Jialing River flow into the main urban area from the southwest and northwest, respectively, and converge at Chaotianmen and then flow to the east, cutting across low mountains or hills along the river and forming valleys, and forming sandbanks or central bars in relatively wide areas after valleys. Divided by river basins, rivers in the main urban area, respectively, belong to the main stream of the Yangtze River and main stream of Jialing River and other small rivers are networks, forming a relatively dense river network with main streams and tributaries in trellised or dendritic pattern. Except for the Yangtze River and Jialing River, the main urban area has 40 primary tributaries with the basin area above 10 km², among which there are 18 tributaries with the basin area above 50 km², including the Bibei River, Liangtan River, Houhe River, Zhuxi River, Baishui Stream, Tiaodeng River, Daxi River, Yipin River, Huaxi River, Kuxi River, Yuxi River, Wubu River, Shuanghe River, Yuzang River, Yulin River, Chaoyang Stream, Chaoyang River and Shuangxi River; and 22 rivers with the basin area above 10 km² and below 50 km², including the Sanxikou, Longtanzi, Jingkou Nanxi Stream, Shuangbei Zhanjia Stream, Panxi River, Tongjia Stream, Qingshui Stream, Zengjia Brook, Jiuqu River, Zhangjia Stream, Sancha River, Mahe Stream, Xipeng Huangjiawan Stream, Huangxi River, Lancao Stream, Shaxi Stream, Wangjiang River, Maoxi Stream, Funiu Stream, Gailan Stream, Taohua Stream and Gulao Stream, and 32 of them are located within the belt freeway.

The main urban area is rich in woodland resources, mainly distributed in areas of the four mountains. The woodland area in the main urban area totals about 1,497 km², accounting for about 27.3% of the planned area. Among them, the woodland within the four mountains is about 928 km², accounting for 62% of total woodland in the main urban area.

The farmland and garden area (cultivated land, garden plot) in the main urban area total about 1,942 km², accounting for about 35.5% of the planned area. The farmland in the main urban area is mainly distributed in north Yubei District and southeast Banan District, with the terraces primarily developed on banks of primary tributaries such as the Maliu River, Yulin River and Ersheng River among the Jinyun Mountain, Zhongliang Mountain (including Longwangdong Mountain), Tongluo Mountain, Mingyue Mountain, Taozidang Mountain and Dongwenquan Mountain.

At present, the main urban area has a total of 285 water surfaces, including 190 small reservoirs (Type II) and 95 other general centralized water surfaces. Four of them are serving as urban drinking water sources, namely, the Nanpeng, Majiagou, Guanyindong and Yinglong reservoirs. Currently, the main urban area has three wetland parks, namely, the Caiyun Lake (national level), Yinglong Lake (national level) and Jiuqu River wetland parks. The reservoirs, lakes and large wetlands are the main spaces for water storage in the water system, which play the roles of flood storage, flood control, regulation of main stream water and improvement of urban local climate, and some reservoirs also play roles in providing drinking water, irrigation water, etc. (Figure 5.2).

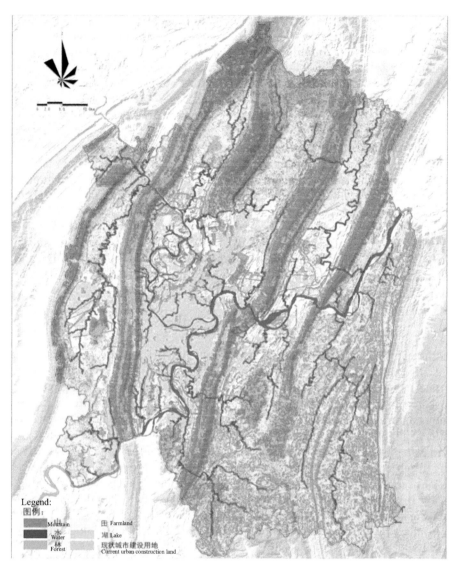

Figure 5.2 Distribution of mountains, rivers, forests, farmland and lakes in Chongqing main urban area.

5.2.2.2 Sponge ecospace pattern

Based on the analysis on natural background of existing mountains, rivers, forests, farmland and lakes, it is planned to construct the sponge city in Chongqing main urban area, forming the overall sponge ecospace pattern with "four mountains, two ridges and forty hills; thousand streams and hundred lakes emptying into Yangtze River and Jialing River; mountains and rivers spreading half the city and green landscape decorating the whole city".

The "four mountains, two ridges and forty hills" mainly refers to four major mountains in the main urban area, namely, the Jinyun Mountain, Zhongliang Mountain (including Longwangdong Mountain), Tongluo Mountain and Mingyue Mountain, and forty mountains in the main urban area, such as the Qiaoping Mountain, Yunzhuan Mountain, Zhaishanping Mountain and Yuntai Mountain. The four mountains, two ridges and forty hills are the spatial carriers for water conservation in the main urban area, mainly reflecting the sponge functions of water "infiltration" and "purification" and also having relevant water "retention" functions to a certain degree.

The "thousand streams and hundred lakes emptying into the Yangtze River and Jialing River" mainly refers to the 2 main streams of the Yangtze River and Jialing River and 40 important primary tributaries, over 2000 secondary and tertiary tributaries, as well as more than 200 lakes and reservoirs. The thousand streams and hundred lakes are the major spatial carriers for water in the main urban area, mainly reflecting the sponge functions of water "storage", "retention" and "utilization".

The "mountains and rivers spreading half the city and green landscape decorating the whole city" mainly refer to the constructed and natural green spaces, such as farmland, woodland, park green spaces, protective greenbelts of river banks and roads. Such spaces are transition areas in the runoff process and important areas for water "purification" and "retention" (Figure 5.3).

5.2.2.3 Sponge function zoning

In the special planning for sponge city construction in the main urban area, it has been specified that the concept of sponge city construction shall be applied to the whole process of Chongqing urban construction in combination with the characteristic of fast stormwater runoff in mountainous area to build an urban water system with virtuous natural circulation, protect the water environment, safeguard the urban water safety, enhance water value and assume the responsibility for water source protection and water and soil conservation in the upper reaches of the Yangtze River. By enhancing the urban waterlogging prevention capacity and improving the quality of new-type urbanization, the goals of "no puddle in case of light rain, no waterlogging in case heavy rain, no black and odorous water and alleviative heat island effect" can be gradually achieved to let the public personally feel the effects of sponge city construction, minimize the urban development and construction's impacts ecological environment and build a healthy and perfect urban water ecosystem.

Combined with analysis on ecospace pattern and analysis on spatial distribution and functions of large sponge bodies in the main urban area, the land within the planned area is divided into four Grade-I sponge functional areas, namely, sponge conservation area, sponge buffer area, sponge improvement area and sponge restoration area.

1. Sponge conservation area.
   The sponge conservation area is about $1,634\,km^2$, accounting for about 29.8% of the planned area. The sponge conservation area mainly refers to the controlled areas of Jinyun Mountain, Zhongliang Mountain, Tongluo Mountain and Mingyue Mountain; continuous mountains such as Longwangdong Mountain, Taozidang Mountain and Dongwenquan Mountain; Zhaishanping Mountain, Qiaoping

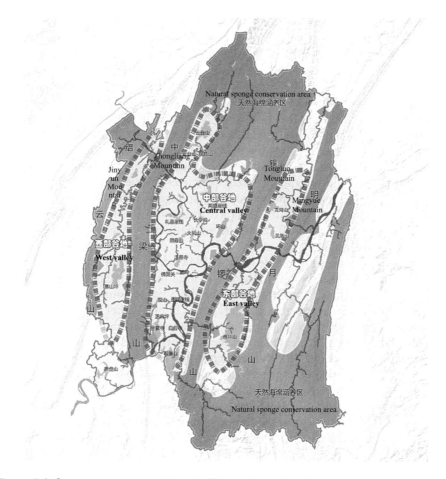

Figure 5.3 Sponge ecospace pattern in Chongqing main urban area.

Mountain, Zhaomu Mountain and other mountains in the main urban area with
very high ecological service functions; and wetland parks such as Caiyun Lake
and Jiuqu River. The sponge conservation area is primary for ecological conservation and protection, where various development and construction activities shall
be strictly controlled; the ecological restoration of areas with water and soil loss or
rocky desertification and the integrated treatment of ecological environment shall
be intensified; sponge facilities, such as impounding reservoir system and stormwater garden, shall be planned and set up in combination with concrete projects
such as forest park, country park and wetland park, so as to improve the conservation function of the sponge spaces.

2. Sponge buffer area.
   The sponge buffer area is about 2,614 km^2, accounting for about 47.8% of the
   planned area. The sponge buffer area is the transition area between sponge conservation area and sponge improvement area, mainly referring to the peripheral farmland and woodland of urban construction areas, where the ecosystem is

unstable and some agricultural non-point source pollution problems exist due to the frequent interference of human activities. The sponge buffer area is primary for ecological protection and buffer, where various development and construction activities shall be controlled; the construction of farmland, forest network and river bank protective greenbelts, as well as the agricultural non-point source pollution treatment shall be intensified; the development of eco-friendly industries such as eco-tourism and eco-agriculture shall be promoted; sponge facilities with regulation and storage functions shall be planned and set up in combination with the spatial distribution of farmland, reservoirs, ponds, depressions, etc. Where the sponge buffer area involves the extension of construction land for future urban development, ecospaces with important sponge functions, such as mountain, water body, pond and woodland, shall be actively protected; with the priority given to the ecospace pattern protection, the sponge system shall be built with green infrastructure and gray infrastructure, respectively, as the main and auxiliary facilities, to promote the LID construction pattern.

3. Sponge improvement area.

The sponge improvement area is about $594\,km^2$, accounting for about 10.8% of the planned area. The sponge improvement area mainly refers to the unconstructed area in the urban planning, including the area of construction land for small towns within the planned area. The sponge improvement area is the core area for the future urban development. The sponge improvement area shall be goal-oriented and focus on the ecological function optimization and construction quality improvement, where, with the ecological civilization construction concept as the core, the priority shall be given to the implementation of blue-green space system; the water system and its green buffer area shall be protected; and the natural ecology protection, urban development and economic input shall be comprehensively balanced, so as to improve the comprehensive benefits of sponge city construction. With green infrastructure and gray infrastructure, respectively, as the main and auxiliary facilities, sponge facilities for the "infiltration, retention, storage, purification, utilization and drainage" shall be reasonably planned and arranged to systematically control runoff and purify water quality from the source, process and end, so as to improve the quality of sponge city construction.

4. Sponge restoration area.

The sponge restoration area is about $631\,km^2$, accounting for about 11.6% of the planned area. The sponge restoration area mainly refers to the existing urban built-up area, including existing regional infrastructure land, etc., where the large-scale hardened pavement, river bank hardening and river re-channeling cause many problems, such as degradation of water system ecological functions, water quality deterioration and serious waterlogging phenomenon. The sponge restoration area shall be problem-oriented and focus on solving problems such as runoff pollution, local ponding and natural infiltration obstruction to restore the ecology. In combination with existing spaces such as park green space, road protective greenbelt, supporting green space, ecological infrastructure such as eco-grass ditch, retention pond and stormwater garden shall be planned and set up; in combination with drainage network and green space and other spaces, initial stormwater facilities shall be planned and set up; the permeable pavement shall be adopted to gradually reconstruct hardened ground; the ecological restoration technologies

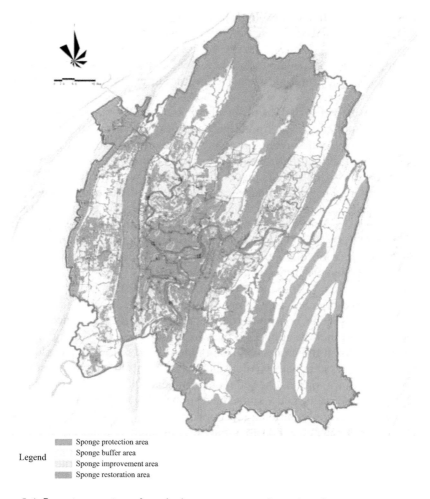

Legend
Sponge protection area
Sponge buffer area
Sponge improvement area
Sponge restoration area

Figure 5.4 Function zoning of grade-I sponge spaces in main urban area.

shall be adopted to gradually restore channelized rivers; the ecological service functions of built-up areas shall be restored as much as possible (Figure 5.4).

5.2.3 Short-term construction planning ideas

On the principle of protection and restoration prior to planning and construction in the sponge city construction, the areas to be protected and restored during the sponge city construction in Chongqing main urban area shall be identified first through eco-logical background investigation. Chongqing main urban area can be divided into five functional areas, namely, sponge ecological protection area, a highly ecologically sensitive area where the development and construction is prohibited to protect the nat-ural ecological pattern; sponge ecological conservation area, an area with important systems of mountain, water, forest, farmland and lake, where the natural ecological

pattern shall be strictly protected; sponge ecological buffer area, an ecologically fragile area, where the artificial ecological restoration is needed to restore to the natural
ecological state; sponge construction pilot area, a sponge city area to be constructed
in the near future, delineated to achieve the construction goal of 20% sponge city by
2020; and sponge construction guidance area, a sponge city area to be constructed
in the long future, delineated to achieve the long-term sponge city construction goal.

1. The old urban area shall be problem-oriented and focus on solving problems such
    as runoff pollution, local ponding and natural infiltration obstruction. On the
    basis of background investigation, the treatment of Chongqing black and odorous
    lakes and black and odorous river segments reported to the MOHURD shall be
    carried out. As for the 267 places vulnerable to waterlogging, LID facilities and
    stormwater management and storage facilities shall be set up in combination with
    the sponge city construction. The complete drainage zones covering the black and
    odorous water bodies and places vulnerable to waterlogging shall be included in
    the sponge city construction demonstration areas, with the treatment results of
    the black and odorous water bodies and places vulnerable to waterlogging as the
    sponge city construction evaluation objectives.
2. New urban area, various industrial parks and large stretches of development zones
    shall be goal-oriented, give priority to the natural ecological background protec-
    tion and reasonably control the development intensity. The concept of sponge city
    construction shall be applied to the whole process of Chongqing urban construc-
    tion, so as to, in the process of urban renewal and renovation, explore to build an
    urban water system with virtuous natural circulation, create an ecological devel-
    opment pattern, protect the water environment, safeguard the urban water safety,
    enhance water value and assume the responsibility for water source protection and
    water and soil conservation in the upper reaches of the Yangtze River. The demon-
    stration zones in new urban area shall be delineated on the principles of centralized
    locations to form continuous areas, covering complete drainage zones and refer-
    ence to concrete quantified goal system. The sponge city construction in new urban
    areas is mainly carried out from the point, line and plane control systems:

Point control system: mainly on the principle of source reduction, it implements the
end control through small urban green spaces and some gray infrastructure easily to
be managed and maintained to reduce runoff pollution, conserve water source, control water and soil loss and reduce peak runoff from the source.

Line control system: mainly on the principle of process control, it builds a systematic sponge system combined with gray and green infrastructure and with equal emphasis on natural ecological functions and artificial engineering measures, which gives
priority to the water system corridors formed by rivers and river vegetation, ecological
corridors formed by urban linear green spaces and linear LID facilities as the main
facilities, supplemented by the optimization and improvement of gray infrastructure
pipe network system, treatment of inland water system and stormwater carrying and
discharge channels, focusing on solving problems of stormwater runoff pollution control, ecological restoration and waterlogging prevention and control.

Plane control system: primarily on the principle of systematic treatment and following
the landscape mechanism with regional green space as the core, it connects the point and

line control systems, focusing on solving problems on aspects such as urban non-point source pollution, total runoff amount control, ecological conservation and protection.

The above three control systems show the sponge city construction measures in an all-around manner from the aspects of source reduction, process control and systematic treatment, reflecting equal emphasis on natural ecological functions and artificial engineering measures, characterized by systematicness, wholeness and integrity.

5.3 Sponge community construction[7][8]

5.3.1 Technical measures for sponge community construction

5.3.1.1 Pollutant intercepting gutter inlet and pollutant intercepting inspection well

The gutter inlet is the "throat" of urban drainage system and main channel for urban non-point source pollutants to enter the water environment, which is of great importance for the water quality control of stormwater runoff. The existing gutter inlets' lack of purification function leads to ineffective source control of stormwater runoff pollution, which is also one of the important reasons why the quality of urban water environment has not improved fundamentally. The setup of pollutant intercepting facilities not only can control the runoff pollutants from the source to purify stormwater before entering the downstream but also can prevent the gutter inlet from blocking and effectively alleviate urban waterlogging, which is especially important in the sponge city construction. The pollutant intercepting inspection well also has a good interception effect on pollutants in the pipe network.

5.3.1.2 Regulating storage pond

As a means of flood retention and stormwater pollution control, the regulation and storage of stormwater are widely used worldwide. Initially, regulating storage ponds were only used to temporarily store excessive stormwater, and natural ponds or depressions were always used to store water. With the people's increasingly deep understanding of storm floods and non-point source pollution, the functions and types of regulating storage ponds are gradually diversified. According to the engineering functions, regulating storage ponds are mainly classified into three types: peak discharge regulation, non-point source pollution control and stormwater utilization. In the construction of communities in mountainous sponge city, these ponds can effectively control the annual runoff discharge and realize utilization of stormwater resources.

5.3.1.3 Stormwater garden

The stormwater garden refers to the engineering facility that can infiltrate and filter stormwater by using soil and plants to purify and retain stormwater at the same time, so as to reduce the runoff. The stormwater gardens in mountainous sponge city communities have many functions such as stormwater management, water purification, stormwater resource utilization and restoration of water cycle. Stormwater gardens shall be connected with stormwater drainage ditches around buildings to collect roof

Figure 5.5 Real picture of stormwater garden.

stormwater and water drained from green spaces and roads in surrounding areas of the gardens, aiming for the retention, slow drainage, evaporation and plant purification of stormwater, contributing to improving the removal rate of pollution load and control rate of total runoff volume (Figure 5.5).

5.3.1.4 Permeable pavement

The permeable pavement system is an important source control technology under the concept of "sponge city". At present, the permeable pavement system has been widely applied in areas such as parks, parking lots, sidewalks, squares and light-load road. Its main functions are to collect, store and treat stormwater runoff, so as to replenish aquifer by infiltration, which is of great significance to improve the overall hydrological regulation and storage function of a city.

5.3.1.5 Stormwater collection and utilization system

The communities in mountainous sponge city utilize the treated stormwater as a non-conventional water resource, which can improve the water resource utilization rate and alleviate the water supply pressure. Meanwhile, the in situ utilization of stormwater resource can effectively reduce urban stormwater runoff, alleviate urban flood peak and reduce pollutants entering the urban water system.

The stormwater runoff is collected through stormwater intercepting well, and enters grit chamber through intercepting pipes, then enters the storage pond after sedimentation treatment. The stormwater is stored and deposited in the storage pond, then pumped up by lift pump to the integrated stormwater treatment equipment, which is composed of an adaptive filter, an integrated dosing device and a UV disinfector and is placed in a buried equipment processing room. The flowchart of the stormwater collection and utilization system is as shown in Figure 5.6.

5.3.1.6 Sunken greenbelt

Sunken greenbelts can collect stormwater runoff generated by surrounding hardened surfaces, and intercept and purify small-flow stormwater runoff by the combined

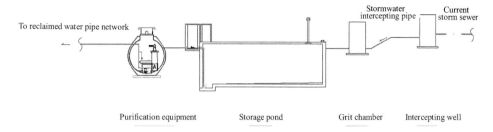

Figure 5.6 Flowchart of stormwater collection and utilization system.

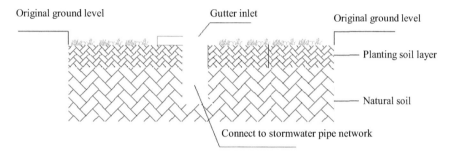

Figure 5.7 Section of sunken greenbelt.

effects of vegetation, soil and microorganisms, while the stormwater exceeding the storage and infiltration capacity will be discharged into the stormwater pipe network via gutter inlets. Sunken greenbelts can reduce the runoff volume and mitigate urban flood disasters. Moreover, the infiltrated stormwater can increase the soil moisture content, thereby reducing the irrigation water consumption of green space and also contributes to the conservation of groundwater. The section of a sunken greenbelt is as shown in Figure 5.7.

5.3.2 Adjust measures to local conditions, collect water in high places for use in low places

Buildings and communities are the starting points of urban stormwater drainage system and important basic units of the sponge city construction. The land for residential quarters accounts for 40%–50% of the total urban construction land, which are important carriers for the sponge city construction. Guided by the sponge city concept, the traditional landscaping pattern of existing buildings and communities shall be changed to accomplish the dual tasks of landscaping renovation and sponge city construction, which is the key to solving the urban water environment problem. Therefore, the application of sponge city concept to buildings and communities is of practical significance to guide the construction of mountainous sponge city. In this section, a community in Chongqing is taken as an example. The effect was evaluated through simulation calculation of the design parameters according to the design scheme for sponge community in Chongqing, and finally the implementation measures for the community's sponge engineering were determined. The renovation design of the engineering construction was guided by the sponge city concept to improve the community's current water environment situation and fully reuse the stormwater.

The community is located in the center of the western area of Chongqing Liangjiang New Area, with the planned construction land area of 18.67 km². As one of the first-batch national "sponge city" construction pilot areas, the community—Yuelai New Town sponge ecological demonstration community—takes "water" as its theme by strictly following the ecological design concept of harmonious and symbiotic natural landforms and community landscape. The sponge city design can effectively improve the stormwater storage and retention capacity through stormwater reuse facilities such as eco-grass ditch, green roof, permeable pavement, pollutant intercepting gutter inlet, stormwater garden and storage pond.

5.3.2.1 Design concept

Due to the intervention of vertical elevation difference and interference of vertical landforms, there are great differences in the community planning and design in mountainous cities and plain cities. The design for communities in mountainous cities is more difficult due to the complex situations of community stormwater drainage and utilization. Combined with the low-lying terrain characteristic of Zhangjiaxi Wetland Park on the northeast of the community, an unobstructed ecological connection corridor between the community and the wetland park was constructed by taking advantage of the mountain terrain and water. The roof and community landscaping and permeable pavement were adopted for the source reduction. Through the intermediate control of storage and purification pond and taking advantage of the terrain, part of the stormwater could flow to the tableland of Zhangjiaxi Park, forming the stormwater wetland landscape and becoming a strong "wetland flood storage" defense, so as to realize the maximum stormwater retention, storage, infiltration, slow release and purification, construct and form the regional "sponge" system.

The sponge city construction flowchart of the community is as shown in Figure 5.8. Combining with the sponge city concept and based on the natural conditions, the construction goals of reaching the specified requirements for pollutant removal rate and annual runoff discharge rate were put forward, with the annual runoff pollutant reduction rate of 25.0% and annual runoff discharge control rate of 45.0% according to the stormwater management and control indicators finally determined after filed investigation.

Figure 5.8 Sponge city construction flowchart of the community.

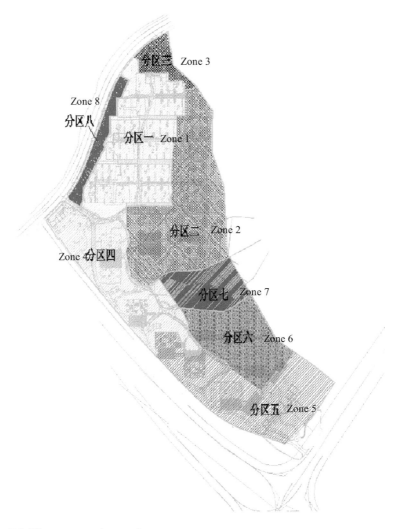

Figure 5.9 The community zoning.

The community was divided into eight zones according to the outdoor ground elevation and stormwater outlets, as shown in Figure 5.9. Based on the analysis on the underlying surface in each zone and according to the existing problems and requirements for the community in the sponge city planning, the design goals were set, and the sponge city construction contents in each zone were determined:

1. Zone 1: pollutant intercepting gutter inlet, pollutant intercepting stormwater inspection well, stormwater landscape pond and stormwater reuse facilities (including concrete storage pond, integrated treatment equipment and reuse system)
2. Zone 2: pollutant intercepting gutter inlet and pollutant intercepting inspection well
3. Zone 3: discharge to end green pace—Zhangjiaxi terraced stormwater wetland

4. Zone 4: stormwater garden, eco-grass ditch, stormwater interception, storage and reuse facilities
5. Zone 5: green roof, stormwater garden, eco-grass ditch, stormwater interception, storage and reuse facilities
6. Zones 6 & 7: stormwater garden, eco-grass ditch.

5.3.2.2 Benefit analysis

The community was adopted with five stormwater gardens, about 500-m reticular eco-grass ditches, "sponge" landscape covering about 40.4% of its area, such as sunken greenbelts, permeable pavements and footpaths without curb, and arranged its "sponge bodies" by following the design concepts of "slow drainage and release" and "source dispersion" and comprehensively adopting technical measures such as water infiltration, retention, storage, purification, utilization and drainage, so as to effectively improve its capacity for water source conservation and purification to realize the natural infiltration, storage and purification of stormwater, forming a perfect community water ecosystem, realizing the perfect integration with the landscape and harmonious development of nature.

The community's pollution load removal rate was 27.4%, higher than the 25.0% required in the regional overall sponge city planning. According to the results of ICM software simulation analysis, the community's annual runoff discharge rate was 39.35%, meeting the requirement of ≤45.5% as specified in the regional overall sponge city planning. Meanwhile, the utilization rate of stormwater resources in the community was 7.83%, with the main stormwater reuse methods mainly including the community's green space irrigation, road washing and water supplement to landscape water bodies. The overflow stormwater close to the Zhangjiaxi side is discharged into the end green space—Zhangjiaxi terraced stormwater wetland, which is discharged after further sedimentation, retention and purification.

To sum up, the implementation of the community's sponge system can effectively reduce the stormwater runoff peak flow. The absorption and interception of eco-grass ditches and stormwater gardens can purify the stormwater before it flows into rivers, weaken the initial scouring effect of stormwater and reduce the initial stormwater overflow. Moreover, the good landscape effect of stormwater gardens and eco-grass ditches is also beneficial to creating a livable environment and improving the community quality (Figure 5.10).

5.3.3 Spot purification and utilization

The sponge city design of another community in Chongqing main urban area was fully combined with the current situations and respects the regional existing functional layout and ecological pattern to construct the community's stormwater management system. Taking the community's actual situation as the basic design condition and solving practical problems as the basic design direction, the design realized the organic coordination between the new engineering system layout and existing drainage system. On the premise of not reducing the drainage capacity of existing drainage system, engineering measures were made according to local conditions, which archived the goals of pollution control, eco-environmental protection and stormwater

Figure 5.10 Effect drawing of a community.

comprehensive utilization while accomplishing the stormwater runoff control indicators. The community's sponge city construction achieved the expected stormwater control effect only by small-scale reconstruction without large-scale modification or change to the existing development facilities.

5.3.3.1 Design idea

The community required the annual runoff pollutant reduction rate $\geq$49.97%, annual runoff discharge rate at each zone $\leq$53.20% and annual total runoff control rate $\geq$79.59%. According to actual conditions, the stormwater management and control indicators for the community were determined as follows: annual runoff pollutant reduction rate $\geq$25% and annual runoff discharge rate $\leq$45.5%.

The design idea of the community is as shown in Figure 5.11: (1) the stormwater drainage sub-catchment and stormwater management zone where the community was located were analyzed according to the water collection scope of the existing pipe network and the terrain features; (2) the plot stormwater management and control goals were determined according to the regional planning; (3) the collection area was divided and the underlying surface was analyzed according to the regional terrain and stormwater pipe network layout; (4) the types and combination schemes of sponge city facilities were preliminarily drafted; (5) the sponge city facility scale were determined and the stormwater management and control effects were determined by model simulation; and (6) the technical and economic evaluation was conducted to finally determine the construction scheme for the mountainous sponge city community.

According to the community's terrain and stormwater pipe network layout, the area was divided into four water collection zones, as shown in Figure 5.12. The areas of Zone 1 to Zone 4 were 15,307.72, 60,349.55, 11,941.67 and 5,826.74 m², respectively.

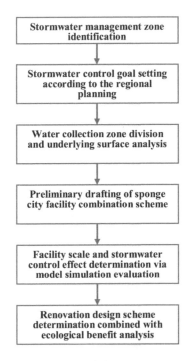

Figure 5.11 Design technology roadmap of the community.

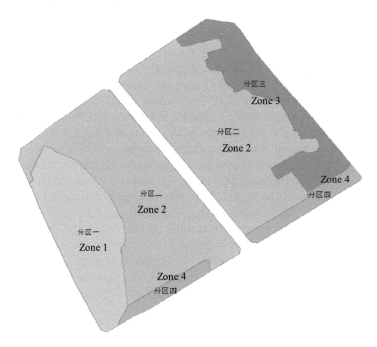

Figure 5.12 Drainage zones of the community.

5.3.3.2 Benefit analysis

The community's runoff discharge rate was 45.05%, slightly higher than the standard of ≤45.50% as put forward in the regional overall sponge city planning and meeting the requirement for annual runoff discharge.

Under the traditional development pattern, the non-point source pollutants of the community are the sum of the pollutants from hardened pavement runoff, green space runoff and roof runoff. After the mountainous sponge city construction, the non-point source pollutants are the sum of the pollutants carried by hard pavement runoff, green space runoff, roof runoff and the remaining pollutants of runoff intercepted by permeable pavements and after being treated in storage ponds. The produced total pollutants were 13.72 t/a (in terms of SS), and the annual runoff pollutant reduction rate was 40.91%, superior to the requirement of 25.0% as stated in the regional overall sponge city planning. The ICM model simulation showed that the community could meet the stormwater management and control indicators as required in the regional planning (Figure 5.13).

5.4 Sponge park construction[2][9][10][11]

This section takes the sponge city reconstruction project of Chongqing Yuelai New Town Exhibition Park as an example. After the overall reconstruction, the Exhibition Park undertakes the functions of stormwater management, storage and reuse in the International Expo area. The functional layout of the original scheme was respected and the current situation was combined with the construction of the Exhibition Park. A stormwater management and control system was built to make the park perform the important functions of stormwater management, storage and reuse in the area after the reconstruction. Besides, on the premise of meeting the park's major functions, various technologies and facilities for stormwater utilization in the park are fully presented by such means as physical display, science popularization and introduction, so as to enhance the public environmental awareness.

5.4.1 Project overview

The Exhibition Park covers an area of 52 hm², with the reconstructed area of 11 hm². The original project site was rocky slopes, with gradients of 1:4–1:2.5 and thin covering

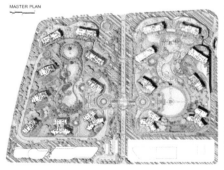

Figure 5.13 Effect drawing of the community.

soil about 1 m thick. The terrain of the park generally rises gradually from the southeast to the northwest, with the highest point located in the middle of the park at an elevation of 324.50 m. Considering the geologic structure and safety factors, the "infiltration" and "retention" should not be adopted for the stormwater management during the construction of the park's LID stormwater system.

In fact, the mountainous cities especially in Chongqing, most of the areas are rocky slopes with thin covering soil. Therefore, during the sponge city reconstruction of such rocky slopes, the excessive increase in soil moisture content may cause the risk of geological disaster. The traditional "infiltration" and "retention" concepts are difficult to be implemented and should not be applied. In this respect, the sponge city reconstruction project of Yuelai New Town Exhibition Park has good demonstration significance (Figures 5.14–5.16).

5.4.2 Design concept

The Exhibition Park is a typical green space park in a mountainous city. During the sponge reconstruction process, the stormwater management measures of "storage", "purification", "utilization" and "drainage" were adopted. Low-lying places and ponds in the park were used to construct stormwater purification, regulation and storage facilities such as eco-grass ditches, stormwater gardens and ecological wetlands. The main runoff direction was identified based on the road and site slope, then stormwater collection facilities were arranged reasonably to realize the regulation, storage, purification and reuse of stormwater.

On the whole, the design concepts of the Exhibition Park were to respect the functional layout of the original scheme, combine with the park's current situation, flexibly select appropriate sponge city measures, construct a stormwater management and

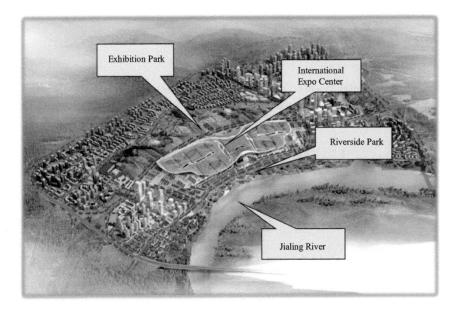

Figure 5.14 Location map of the exhibition park.

Collection area: 7.7 hm^2
Water body area: 210 m^2
Water storage volume: 370 m^3

Collection area: 10.7 hm^2
Water body area: 6,850 m^2
Water storage volume: 6,850 m^3

Collection area: 4.4 hm^2
Water body area:1,400 m^2
Water storage volume: 470 m^3

Figure 5.15 Aerial view of the exhibition park.

control system on the premise of ensuring the park's geological structure safety and make the park perform the functions of stormwater management, storage and reuse in the International Expo area after the reconstruction, so as to achieve the following goals:

1. Control the stormwater pollution and realize the stormwater reuse. The stormwater management, storage, purification and reuse can be realized through optimizing the stormwater runoff directions in surrounding areas and within the park, so as to achieve the goals of reducing the loss of stormwater resource, controlling water pollution and alleviating drainage pressure in surrounding areas.
2. Improve the park's ecological environment and perform the park's function as a position for science popularization. On the premise of meeting the park's major functions, relevant technologies and facilities for stormwater utilization in the park can fully presented by such means as physical display, science popularization and introduction, so as to enhance the public awareness of water resource management (Figure 5.17).

The integrated mode of "stormwater landscape pond + water collection module" was adopted to realize the purification, retention, storage and reuse of stormwater. After the field investigation, the sponge city reconstruction idea of the Exhibition Park was determined as follows:

1. The "stormwater landscape pond + water collection module" were added in the small theater at the park's north end

Figure 5.16 Slope analysis of the exhibition park.

2. The "stormwater landscape pond+water collection module" were added in the Sculpture Square at the park's south end

3. The existing mountaintop pond was built into a "stormwater landscape pond"

4. The "mountaintop landscape pond" and the "Sculpture Square landscape pond" were connected through the lifting of water pumps and transport of eco-grass ditches, forming a landscape water group

5. One water landscape was added in the Phase-II square of the conference exhibition hall, with the water supplied by the existing stormwater reuse facilities. Meanwhile, the water supply capacity of existing stormwater reuse facilities was rechecked in the scheme (Figure 5.18).

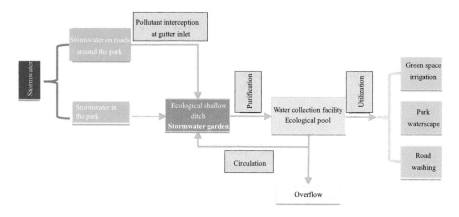

Figure 5.17 Sponge city reconstruction idea of the exhibition park.

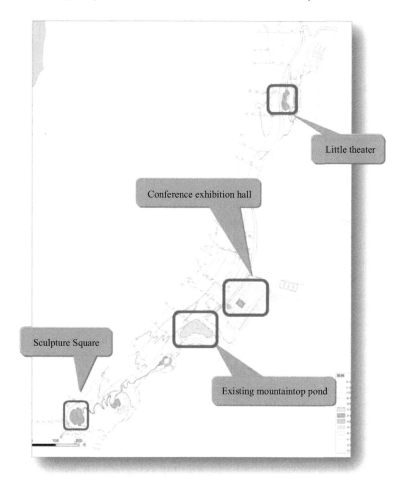

Figure 5.18 Distribution of reconstruction measures for the exhibition park.

5.4.3 Technical measures[14]

According to the geographical distribution pattern, the Exhibition Park was divided into two north and south parks for the implementation of the sponge city reconstruction project, with the north part covering the little theater, and the south part covering the Sculpture Square, mountaintop pond and conference exhibition hall.

5.4.3.1 "Stormwater landscape pond + water collection module" in the small theater at the park's north end

This area was adopted with the reconstruction mode of "stormwater landscape pond + water collection module", namely to collect the stormwater on half breadth of Xuetang Road nearing the north part, plot 3 of Sunac project and green space around the little theater to the stormwater landscape pond in the middle by such collection means as interception by existing stormwater pipe network and interception and transport by eco-grass ditches. The stormwater reuse could be realized in combination with the stormwater reuse system, with the water conservation rate reaching 17%–27% (Figures 5.19 and 5.20).

5.4.3.2 Existing mountaintop "stormwater landscape pond"

The mountaintop stormwater landscape pond collects the stormwater from the park green space around the mountaintop pond, conference exhibition hall and part of Diecai Mountain project plot to the mountaintop stormwater landscape pond by such collection means as natural water collection, interception by existing stormwater pipe network and jacked pipes to the mountaintop pond. The water conservation rate could reach 27.61%–38.13% by use of the stormwater reuse system (Figures 5.21 and 5.22).

5.4.3.3 "Stormwater landscape pond + water collection module" in the Sculpture Square at the park's south end

The Sculpture Square at the park's south end was adopted with the reconstruction mode of "stormwater landscape pond + water collection module" to collect the stormwater from the square to the stormwater landscape pond by the interception and transport of newly built eco-grass ditches. The water conservation rate could reach 8.33%–12.86% by use of the stormwater reuse system.

The stormwater pond in the south part is located at the south entrance of the Exhibition Park. In the project, a stormwater pond was built on the low-lying area, covering an area of $4,700 \text{m}^2$ and with an average water storage depth of 0.7 m. The water collection module was installed below the stormwater pond, with the water collection module volume of 470m^3, which is mainly for collecting the park green space stormwater in the south part, with a collection area of $44,400 \text{m}^2$. The collected stormwater will be used for the park's green space irrigation and road washing through the lift pump station after being purified by a variety of aquatic plants in the pond and being filtered and disinfected by the buried integrated treatment equipment. The combination between stormwater pond and winding eco-grass ditches not only meets the requirements for the retention, purification and collection of stormwater in the park but also forms the beautiful scenery in the park (Figures 5.23 and 5.24).

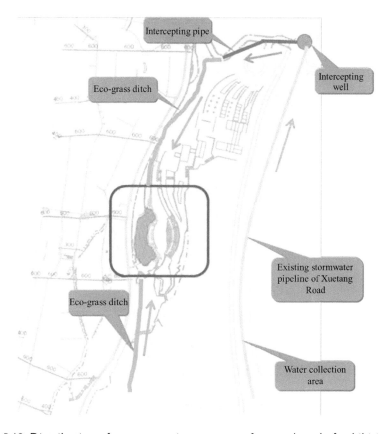

Figure 5.19 Distribution of reconstruction measures for north end of exhibition park.

Figure 5.20 Stormwater pond in the exhibition park.

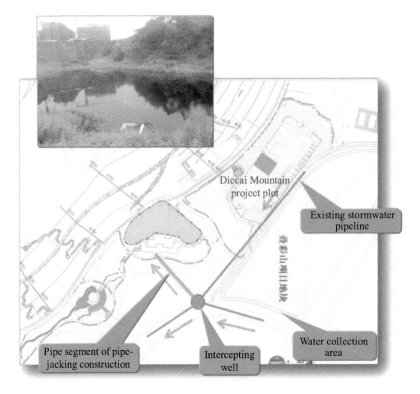

Figure 5.21 Distribution of reconstruction measures for mountaintop stormwater landscape in the exhibition park.

Figure 5.22 Aerial view of stormwater pond in the exhibition park.

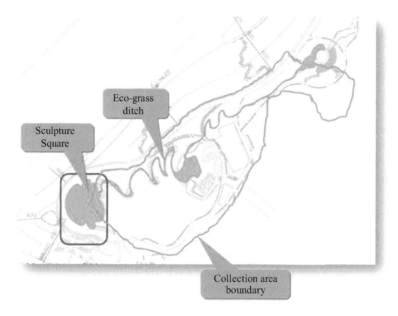

Figure 5.23 Distribution of reconstruction measures for south end of exhibition park.

Figure 5.24 Eco-grass ditch in the exhibition park.

5.4.3.4 Design of connected "stormwater landscape pond + water collection module" in the Sculpture Square and mountaintop pond

The stormwater in the mountaintop pond is lifted to the eco-grass ditches by lift pump, which passes through the Yuantai Square and Lamo Square, landscape water bodies designed in the squares and finally is transferred to the Sculpture Square. Meanwhile, water in the water collection module of Sculpture Square can supplement the mountaintop pond by the water pump lifting, so as to archive the connection between the two and realize the benign cycle of water bodies (Figure 5.25 and 5.26).

5.4.4 Benefit evaluation

5.4.4.1 Achieving goals of stormwater pollution control and stormwater reuse

The stormwater management, storage, purification and reuse can be realized through organizing the stormwater runoff directions in surrounding areas and within the park, so as to achieve the goals of reducing the loss of stormwater resource, controlling water pollution and alleviating drainage pressure in surrounding areas.

5.4.4.2 Improving the park's ecological environment and performing the park's function as a position for science popularization

On the premise of meeting the park's major functions, various technologies and facilities for stormwater utilization in the park are fully presented by such means as physical display, science popularization and introduction, so as to enhance the public environmental awareness (Figure 5.27).

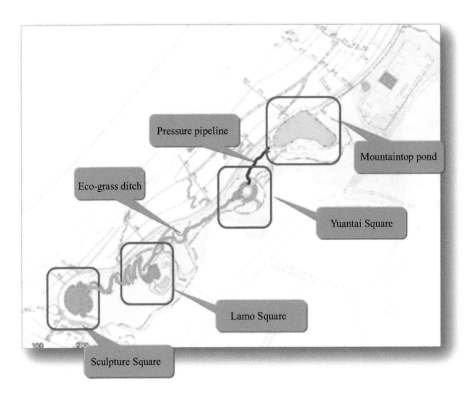

Figure 5.25 Distribution of reconstruction measures for Sculpture Square in the exhibition park.

Figure 5.26 Stormwater pond in Sculpture Square of the exhibition park.

Figure 5.27 Effect drawing of the exhibition park.

Chapter 6

Practice of integrated water environment treatment

The upper reaches of the main stream of the Yangtze River are from the source to Yichang, 4,504 km long, accounting for 70.4% of the total length of Yangtze river, with the basin area of 1 million km^2. The middle reaches are from Yichang to Hukou, 955 km long, with the basin area of 680,000 km^2. The Three Gorges Reservoir Region is located in the middle and upper reaches of the Yangtze River, including Chongqing and Hubei sections. Chongqing section of the Three Gorges Reservoir Region is criss-crossed by many rivers, dense with water networks and abundant in water resources. The main stream of the Yangtze River flows across Chongqing from southwest to northeast. Five major tributaries, namely Jialing River, Qujiang River, Fujiang River, Wujiang River and Daning River, and hundreds of middle and small rivers empty into the Yangtze River, forming an approximately centripetal water system. In addition to the Yangtze River and its major tributaries Jialing River and Wujiang River, there are ten rivers with a basin area above 3,000 km^2. Main rivers in Chongqing include the Yangtze River, Jialing River, Wujiang River, Fujiang River, Qujiang River, Qijiang River, Yulin River, Longxi River, Laixi River, Furong River, Anju River, Panxi River, etc., most of which belong to the Yangtze River system.

The Panxi River is a tributary of the Jialing River. Due to imperfect previous planning and inadequate construction consideration, the drainage network was not perfect, with a small amount of sewage directly discharged into the river. Besides, some sewage pipe networks within the basin were adopted with combined system. In the rainy season, part of the combined sewage overflowed into the combined pipes. Meanwhile, due to the insufficient understanding of stormwater management, the initial stormwater was not properly managed, resulting in the water quality of some water bodies inferior to Class V, or even the black and odorous phenomena of some lakes and reservoirs. In order to fundamentally change such situation, the government started the integrated basin treatment in early 2016. Therefore, the Panxi River is selected as a typical practice case of integrated water environment management.

6.1 Overview of environment in the Panxi River basin

The Panxi River in Chongqing main urban area originates from Liangjiang New Area, flows through Yubei District, and finally empties into the Jialing River in Jiangbei District, with a total basin area about 28.25 km^2. The Panxi River basin flows across Jiangbei District and Yubei District of Chongqing, and there are ten lakes and reservoirs along the river, namely, Cuiwei Lake, Bayi Reservoir, Qingnian Reservoir,

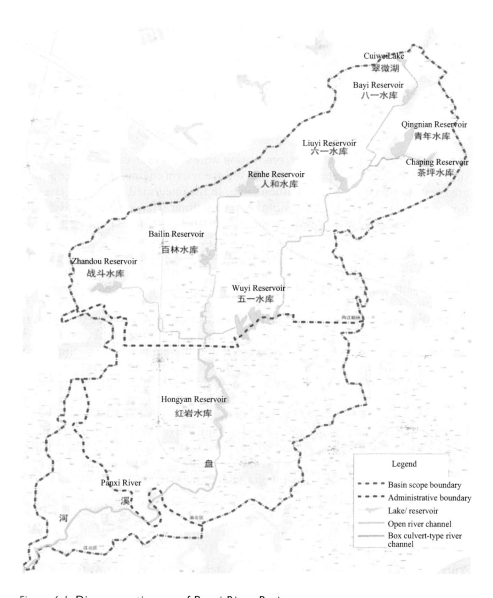

Figure 6.1 Diagrammatic map of Panxi River Basin.

Chaping Reservoir, Liuyi Reservoir, Renhe Reservoir, Bailin Reservoir, Wuyi Reservoir, Zhandou Reservoir and Hongyan Reservoir. The Panxi River basin is as shown in Figure 6.1.

6.1.1 Meteorological information of the basin

The Panxi River basin is located in a subtropical humid monsoon climate zone. According to the statistics over 45 years, the multi-year average temperature is at 18.1°C, with the extreme maximum and minimum temperatures at 40.0°C and −3.8°C, respectively, multi-year average sunshine duration of 1,325 hours, annual average sunshine of 1,341.1 hours, multi-year average precipitation of 1,078 mm, multi-year average evaporation of 1,011.1 mm, multi-year average relative humidity of 80%, multi-year average wind velocity of 1.0 m/s, multi-year average 10-minute maximum wind velocity of 15 m/s, average frost-free period of 350 d and average fog days of 69.3 d.

6.1.2 Renovation works before systematic integrated treatment

In order to treat the Panxi River, renovation works had been continuously carried out since 2008, such as the separate rain and sewage system engineering in Renhe old town area with an investment of 14 million yuan. In the integrated secondary river renovation works carried out in 2010, more than 63 million yuan was successively invested in implementing such measures as the pipe network inspection and restoration, domestic pollution source treatment, industrial pollution treatment, stopping livestock and poultry breeding along the basin, river dredging and ecological restoration, completing the pipe reconstruction of 4,750 m, river dredging of 3 km, garbage cleaning of 12 t and pipe inspection of 4,448 m. The major treatment projects included: reconstruction of Chongqing Municipal Administration Committee pipe network section in Huangshan Avenue, reconstruction of Jinke Hotel and Tianlaicheng Community pipe network section in Huangshan Avenue, renovation of Renhe Avenue drainage network section, improvement of tertiary pipe network in Renjialiangzi area, construction of pollutant intercepting main pipe of Zhandou Reservoir and dredging of Liuyi Reservoir. In 2013, the integrated treatment of municipal drainage facilities in the Panxi River Basin in Liangjiang New Area was carried out; in 2016, the integrated Xiongjiagou treatment project was carried out. After continuous treatment, water quality of the Panxi River was constantly improved but still did not reach the evaluation criteria, and the basin still needed systematic integrated treatment.

In addition to the above projects, the following projects are those that had been completed before the systematic integrated treatment started in 2017.

6.1.2.1 Completed pipe network and pollutant interception projects

6.1.2.2 Completed endogenous source treatment works

Yubei District completed the integrated water environment renovation of Hongyan Reservoir on October 10, 2015, with a total investment of 514,000 yuan (Table 6.1). The current situation of the reservoir after the integrated renovation is as shown in Figure 6.2. The major project contents included rehabilitation and expansion of constructed wetland, setup of reaeration device and building of shallow water ecosystem. The engineering quantity list is as shown in Table 6.2.

6.1.2.3 Other completed treatment projects

Chongqing has carried out the integrated treatment of lakes and reservoirs since 2010 and has achieved some effects. Table 6.3 shows the information about the treatment of lakes and reservoirs in the basin.

Table 6.1 Summary of Panxi River pipe network projects implemented in 2010

S/N	Engineering content	Construction status	Responsible authority
1	Reconstruction of Chongqing Municipal Administration Committee pipe network section in Huangshan Avenue	Completed	Liangjiang New Area
2	Reconstruction of Jinke Hotel and Tianlaicheng Community pipe network section in Huangshan Avenue	Completed	Liangjiang New Area
3	Renovation of Renhe Avenue drainage network section	Completed	Liangjiang New Area
4	Improvement of tertiary pipe network in Renjialiangzi area	Completed	Liangjiang New Area
5	Construction of pollutant intercepting main pipe of Zhandou Reservoir	Completed	Liangjiang New Area
6	Dredging of Liuyi Reservoir	Completed	Liangjiang New Area
7	Reconstruction of sewage pipes in Fortune Center Branch, outside Lake Paradise (Wuxiren Yepi, a night snack shop), Gumufeng Interchange, Hongye Road, etc.	Completed	Liangjiang New Area

Figure 6.2 Current situation of Hongyan Reservoir after integrated renovation.

Table 6.2 Summary of Hongyan Reservoir renovation works implemented in 2015

Category	Specific work	Engineering measure name	Main content	Fund (10^4 yuan)
Non-point source	Initial stormwater purification, treatment and covering of bare ground	Slope landscaping and vegetation planting	Planting shrubs and turf of 250 m^2, and Ophiopogon Japonicus of 6,800 m^2	16.2
		Replanting in bare soil area	The District Park Management Center has replanted for an area of 500 m^2	0.5
		Exposed pipe network covering	Exposed lakeside pipe network with an area of 100 m^2 has been covered	0.1
		Flotsam cleaning	The District Park Management Center has cleaned up the flotsam and carries out periodic cleaning	0.3
	Sewage outfall renovation	Sewage outfall and overflow outlet renovation	The treatment on two lakeside discharge overflow outlets has been carried out	12
	Domestic waste treatment	Waste cleaning and transportation	The District Park Management Center has cleaned and transported the waste and has specially assigned persons for periodic cleaning and transportation	1.5
Endogenous	Direct treatment of water body	Establishment of reaeration facilities	Three aerators have been set up	3.1
	Ecological restoration	Shallow water ecosystem building	One shallow water ecosystem has been built	7.5
		Waterfront wetland	Planting of aquatic plants	10.2

6.1.3 Water environment before systematic integrated treatment

According to the monitoring data of 2015–2016 provided by Liangjiang New Area, Yubei District and Jiangbei District Environmental Protection Bureaus, the concentration ranges of COD, TP, NH$_3$-N and TN in lakes, rivers and river segments within the basin were 7–45, 0.03–2.05, 0.078–15.5 and 1.29–19.1 mg/L, respectively. The pollutant indications of the lakes, rivers and river segments did not meet the quality standards for Class V surface water as specified in *Environmental Quality Standards for Surface Water* (GB 3838-2002), and the water quality of Liuyi Reservoir, Wuyi Reservoir, Panxi River segments and Hongyan Reservoir was perennially interior to Class V (Table 6.4).

The appearance of water bodies is as shown in Figures 6.3–6.6.

Table 6.3 Relevant treatment works for lakes and rivers along the Panxi River

Admin-istrative district	Lake/ reservoir name	Engineering content										Fund (10⁴ yuan)
		Water supple-ment (t)	Pipe network reconstruc-tion/restora-tion (m)	Construc-tion of new pipe network (m)	Water circulating & aeration system (Item)	Waste cleaning & transpor-tation (t)	Shallow water eco-system (m²)	Dredging (t)	Ecological restora-tion (m²)	Biomanipu-lation (No.)	Stormwater interception & filtration ditch (m)	
Liangjiang New Area	Bayi		357				500	13,000	2,000		1,300	302.11
	Qingnian	26,640					5,000	780				127.65
	Chaping					2,200	1,000		200			26.8
	Bailin						8,110					330
	Liuyi		1,134			26	6,591	1,150	7,374			173.89
	Wuyi	44		1,200	40	550	6,000			160,000	16,000	2.227
	Zhandou		5,000			10	1,000	1,000	100	10,000		458.14
	Total	26,684	6,491	1,200	40	2,786	28,201	15,930	9,674	170,000	17,300	3,645.59
Yubei District	Hongyan		245		3		100		2,300		89	62.6
Total		26,684	6,736	1,200	43	2,786	28,301	15,930	11,974	170,000	17,389	3,708.19

Table 6.4 Statistics of water quality monitoring data of lakes and rivers in the Panxi River Basin unit: mg/L

Lake/reservoir name	Sampling date	COD	NH₃-N	TN	TP
Panxi River	January 4, 2015	16	6.98	/	1.12
	February 2, 2015	48.9	9.32	/	1.08
	March 2, 2015	74.5	6.72	/	1.38
	April 1, 2015	22.5	2.61	/	0.539
	May 4, 2015	35.4	3.54	/	0.246
	June 1, 2015	31.5	2.67	/	0.412
	July 1, 2015	17.9	2.66	/	0.35
	August 3, 2015	23.5	8.63	/	1
	September 1, 2015	27.4	1.95	/	0.367
	January 4, 2016	62.7	5.67	8.71	0.648
	January 5, 2016	164	4.3	13.6	/
	January 6, 2016	20.6	3.34	6.69	/
	January 7, 2016	21.9	3.9	8.56	0.576
	January 8, 2016	36.8	3.7	10.5	/
	January 9, 2016	64.3	4.13	8.12	/
	January 10, 2016	26.9	3.62	7.96	/
	January 11, 2016	24.4	3.15	5.82	0.488
	January 12, 2016	28.8	3.34	7.58	/
	January 13, 2016	30.4	4.01	8.86	/
	January 14, 2016	24.9	3.5	7.66	0.492
	January 15, 2016	21	3.31	6.79	/
	January 16, 2016	29.5	3.4	6.72	/
	January 17, 2016	21.2	3.86	7.27	/
Panxi River	January 18, 2016	25.3	3.81	7.07	0.52
	January 19, 2016	31.4	3.96	7.98	/
	January 20, 2016	25.4	3.21	7.29	/
	January 21, 2016	31	3.34	5.98	0.46
	January 22, 2016	34.7	3.55	8.36	/
	January 23, 2016	81.3	8.39	13.6	/
	January 24, 2016	68.3	8.31	14.5	/
	January 25, 2016	29	4.19	9.76	0.624
	January 26, 2016	29.8	4.73	8.22	/
	January 27, 2016	42.7	3.47	8.08	/
	January 28, 2016	22	7.84	6.12	0.852
	January 29, 2016	26.6	3.59	8.6	/
	January 30, 2016	23.5	3.44	6.73	/
	January 31, 2016	42.3	4.46	8.46	/
	February 1, 2016	35.1	3.51	7.86	0.52
	February 2, 2016	26.4	2.85	6.65	/
	February 3, 2016	22.1	2.65	5.8	/
	February 4, 2016	45.6	3.54	7.47	0.576
Bayi Reservoir	May 28, 2015	24.2	/	2.11	0.094
	November 26, 2014	12.9	/	5.31	0.204
	January 21, 2015	11.8	/	3.63	0
		7.36	/	5.6	0.031
	April 1, 2015	19.4	/	1.32	0.457
	April 8, 2015	19.4	/	1.32	0.457
Bayi Reservoir	August 11, 2015	27.5	/	3.31	0.076
	May 17, 2016	34.4	/	5.92	0.19
Bailin Reservoir	April 1, 2015	16.3	/	3.6	0.564
	May 28, 2015	31.3	/	3.65	0.211
	November 26, 2014	15	/	3.7	0.385

(Continued)

Table 6.4 (Continued) Statistics of water quality monitoring data of lakes and rivers in the Panxi River Basin unit: mg/L

Lake/reservoir name	Sampling date	COD	NH₃-N	TN	TP
Chaping Reservoir	May 28, 2015	31.8	/	1.33	0.05
	November 26, 2014	15.9	/	2.69	0.129
	April 8, 2015	52.9	/	2.93	0.782
	August 11, 2015	22.1	/	1.35	0.06
	May 17, 2016	32	/	1.24	0.054
Hongyan Reservoir	November 4, 2014	24.6	/	3.84	0.095
	February 10, 2015	22.9	2.58	8.49	0.859
	May 11, 2015	19.6	2.09	4.25	0.343
	August 4, 2015	39.9	2.48	4.87	0.479
	November 9, 2015	20	6.36	7.21	0.75
	May 5, 2016	22.1	/	7.43	0.193
Liuyi Reservoir	May 28, 2015	45.2	/	2.52	0.298
	November 26, 2014	9.5	/	3.78	0.58
	April 1, 2015	14.6	/	1.2	0.196
	August 11, 2015	26	/	2.92	0.259
	May 17, 2016	34.7	/	1.51	0.102
Qingnian Reservoir	May 28, 2015	35.3	/	1.54	0.065
	November 26, 2014	33	/	1.95	0.129
	April 8, 2015	30.5	/	5.76	0.192
Qingnian Reservoir	August 11, 2015	22.3	/	1.88	0.068
	August 17, 2016	35.3	/	1.3	0.057
Renhe Reservoir	August 28, 2015	29.9	/	1.24	0.06
	April 1, 2015	17.6	/	1.42	0.113
	August 11, 2015	25.6	/	1.44	0.054
Wuyi Reservoir	November 4, 2014	17.2	/	2.31	0.133
	November 9, 2015	18.3	/	0	0
	May 23, 2016	21.8	0.258	1.96	0.064
		23.36	0.252	1.82	0.088
		24.92	0.228	2.04	0.064
		23.36	0.258	1.96	0.064
		28.04	0.914	2.26	0.144
		24.92	0.692	2.14	0.104
	June 1, 2016	15	0.078	1.29	0.058
		17.6	0.278	1.81	0.114
		22.7	1.27	3.57	0.144
		16.6	1.44	3.71	0.18
		42.3	3.98	7.78	0.5
		25.5	2.8	5.82	0.319
	June 2, 2016	14.6	1	2.78	0.133
		15.8	1.2	3.29	0.147
		18.2	1.47	3.86	0.19
		16.4	1.27	3.53	0.143
		28.9	3.85	6.24	0.333
Wuyi Reservoir	June 2, 2016	18.6	2.46	4.92	0.172
	June 6, 2016	27.1	0.993	3.64	0.139
		20.6	1.28	3.86	0.122
		27.8	2.67	4.53	0.322
		25.2	2.41	5.39	0.257
		196	15.5	19.1	2.05
		29.9	2.91	6.1	1.7
Zhandou Reservoir	April 8, 2015	34	/	3.23	0.157
	May 17, 2016	24.2	/	1.46	0.156
	November 26, 2014	0	/	5.25	0.371

Figure 6.3 Photos of Liuyi Reservoir.

(a) Box culvert outlet   (b) Reservoir surface

Figure 6.4 Photos of Wuyi Reservoir. (a) Box culvert outlet. (b) Reservoir surface.

Figure 6 5 Photos of Panxi River segments.

Figure 6.6 Photos of Hongyan Reservoir.

6.2 Background investigation and main problems of pollution in the Panxi River basin[3]

In order to find out the actual situation in the Panxi River basin and put forward pertinent and systematic integrated treatment measures, a background investigation on the basin was started in 2017. A comprehensive investigation on pollution sources in the basin was carried out by such means as data collection and analysis, field investigation, pollution source investigation and water quality sampling analysis.

6.2.1 Pollution source investigation

Generally, lake pollution sources can be divided into exogenous pollution and endogenous pollution sources. The lake endogenous pollution mainly refers to the phenomenon that nutrients entering lakes gradually deposit on the sediment surfaces of lakes under various physical, chemical and biological effects. The lake exogenous pollution includes point and non-point source pollution.

6.2.1.1 Investigation on current situation of exogenous pollution

Main sources of exogenous pollution:

1. Point source pollution.
   The separate rain and sewage systems were partly adopted in areas such as Huangshan Avenue, Crystal Central, Evergrande Palace, Lake Paradise, Jinkai Avenue, Palm Springs Community, Hongmian Avenue, Jinxiu Shanzhuang, Jinlong Road, Songqiao Road, Donghe Chuntian, Heji Primary School and Panxi Road, but the separation was not complete and the phenomenon of direct discharge of domestic sewage still existed. On both banks of Panxi River, there were still many intercepting wells in incompletely separate systems with low interception ratios. As a result, overflows occurred frequently in rainy seasons, leading to the mixed stormwater and domestic sewage entering lakes, reservoirs and rivers, causing the overflow pollution of Panxi River. Due to loose management on construction wastewater, a large amount of construction wastewater containing sands was discharged into Hongyan Reservoir and the downstream river. The major sources of point source pollution were as shown in Table 6.5, and the distribution locations were as shown in Figures 6.7–6.9.
   The distribution of exogenous pollution loads was as shown in Figure 6.10.
2. Point source pollution loads of lakes and rivers in the basin.
   The sewage receiving range of each lake or reservoir was defined according to the drainage network and distribution of point sources as shown in Figure 6.10, and the calculated volume of sewage directly discharged was as shown in Table 6.6. Relevant data in the table were calculated based on the following values: population density of 15,000 persons per km^2, integrated water quota per unit population of 420 L/(cap/d), pollutant producing coefficient of 0.85, pipe network collection rate of 0.98, groundwater infiltration rate of 1.05 and daily variation coefficient of 1.2.
   The monitoring and analysis of sewage water quality in Chongqing main urban area showed that the average concentrations of domestic sewage pollutants were shown in Table 6.7.
   Based on the sewage receiving range in the basin and pollutant concentrations of domestic sewage, the loads of point source pollutants in relevant lakes, reservoirs and rivers could be predicted, as listed in Table 6.8.
3. Non-point source pollution.
   According to relevant research data, the non-point source pollution such as urban stormwater runoff pollution accounted for more than 15% of the water body

Table 6.5 Results of investigations on major point source pollution sources of Panxi River

Pollution source no.	Local administrative district	Sewage outfall type	Existing situation
LJW1	Liangjiang New Area	Outfall for mixed stormwater and sewage	A large amount of mixed sewage overflowed through the overflow outlets in rainy seasons, polluting the river course between Wuyi Reservoir and Liuyi Reservoir, and finally flowing into Wuyi Reservoir
LJW2	Liangjiang New Area	Outfall for direct discharge of domestic sewage	A large amount of domestic sewage was discharged from Jinke Tianlaicheng Meishe and its surrounding communities. It was one of the main sewage outfalls to Panxi River (which had been treated at the beginning of 2017 and was a small amount of sewage is still discharged)
LJW3	Liangjiang New Area	Outfall for direct discharge of domestic sewage	A large amount of domestic sewage was discharged from Chongqing Fire Brigade and its surrounding communities. It was one of the main sewage outfalls to Panxi River
LJW4	Liangjiang New Area	Outfall for mixed sewage and construction wastewater	A large amount of light rail station construction sandy wastewater was discharged, and the mixed sewage was discharged from Xiongjiagou tributary (treatment project being carried out)
LJW5	Liangjiang New Area	Outfall for direct discharge of domestic sewage	A small amount of domestic sewage was discharged from areas behind Liangjiang New Area Planning and Exhibition Hall
LJW6	Liangjiang New Area	Outfall for direct discharge of domestic sewage	A small amount of domestic sewage was discharged from areas behind Liangjiang New Area Planning and Exhibition Hall
LJW7	Liangjiang New Area	Outfall for direct discharge of domestic sewage	A large amount of domestic sewage was discharged from the basin of Zhandou Reservoir tributary
LJW8	Liangjiang New Area	Outfall for direct discharge of domestic sewage	A small amount of domestic sewage was discharged from basketball court of Dongbu Park
YBW1	Yubei District	Outfall for mixed stormwater and sewage	It was a stormwater outfall of Tianyi Xincheng Community, but part of domestic sewage was mixed into the outfall and discharged into the river
YBW2	Yubei District	Outfall for mixed stormwater and sewage	Four lakeside inspection wells were connected to Hongyan Reservoir, with relatively high water levels currently, very likely to cause overflow in rainy seasons

(Continued)

Table 6.5 (Continued) Results of investigations on major point source pollution sources of Panxi River

Pollution source no.	Local administrative district	Sewage outfall type	Existing situation
YBW3	Yubei District	Outfall for mixed stormwater and sewage	As for the sewage outfall and two DN500 double-wall corrugated drainage pipes of North International Center Community, the domestic sewage was mixed with stormwater and directly discharged into rivers
YBW4	Yubei District	Outfall for direct discharge of domestic sewage	A small amount of domestic sewage (probably kitchen wastewater from Beijiang Roast Whole Lamb and domestic sewage from Dongheng Longdu) was directly discharged
YBW5	Yubei District	Outfall for construction wastewater	A large amount of sandy construction wastewater was discharged
YBW6	Yubei District	Outfall for mixed stormwater and sewage	Two inspection wells were connected to Panxi River, with relatively high water levels currently, very likely to cause overflow in rainy seasons
YBW7	Yubei District	Outfall for direct discharge of domestic sewage	A small amount of domestic sewage was directly discharged into rivers through outfalls around Yubei Huayuan Primary School, and the drainage pipes within river levees slightly leaked
YBW8	Yubei District	Outfall for direct discharge of domestic sewage	A small amount of domestic sewage was directly discharged from areas behind the Songshilu Community
JBW1	Jiangbei District	Outfall for direct discharge of domestic sewage	A large amount of domestic sewage was directly discharged from Daqing Community
JBW2	Jiangbei District	Outfall for mixed stormwater and sewage	Due to the small interception ratios of intercepting wells in incompletely separate system, the intercepting wells were nearly full in dry seasons, with a large number of annual overflow times. The sewage from communities near Yuxiangmen and Songshi Avenue was discharged into Panxi River in rainy seasons
JBW3	Jiangbei District	Outfall for mixed stormwater and sewage	A small amount of mixed stormwater and sewage discharged from a plant of Changan Automobile
JBW4	Jiangbei District	Outfall for direct discharge of domestic sewage	Sewage outfall of Shimen Community

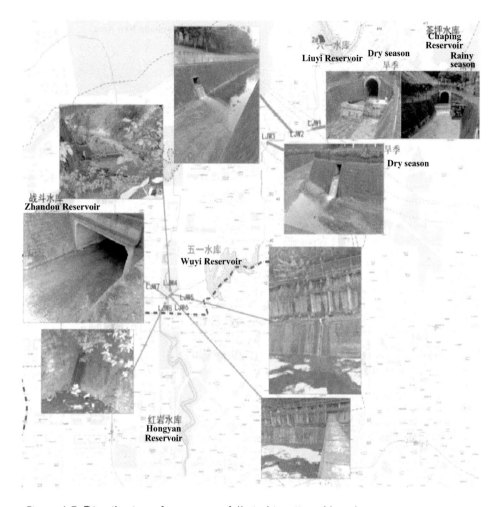

Figure 6.7 Distribution of sewage outfalls in Liangjiang New Area.

pollution load, and the stormwater runoff had become an important factor in the water quality deterioration. The background investigation showed that the non-point sources in the surroundings of Panxi River came from the drainage box culvert service area and stormwater runoff pollution around the river and lakes.

According to the results of researches carried out by Chinese Academy of Sciences, Southwest University and Chongqing University on the water quality of stormwater runoff from different underlying surfaces in Chongqing, such as *Water Quality Characteristics of Rainfall Runoff from Different Roofs in Urban Area, Characteristics and Control of Urban Rainfall-runoff in Mountainous City, Characterization of Stormwater Runoff Pollution in Mountain City, Characteristics of Runoff Pollution on Urban Pavements with Different Materials in Chongqing, Characteristics of Runoff from Different Material Roofs in Chongqing Urban Area* and

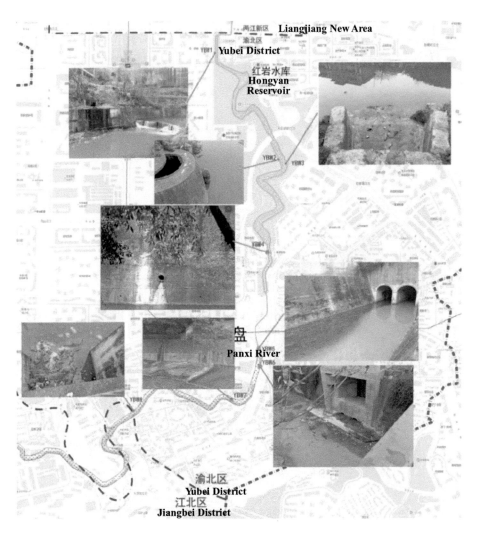

Figure 6.8 Distribution of sewage outfalls in Yubei District.

Study on Rainfall Runoff Pollution and Control at Different Underlying Surfaces in Urban Residential Area of Chongqing, the pollutant indicators for stormwater runoff in Chongqing were obtained (Table 6.9).

Calculated on the basis of the area proportions of road, green space, roof and hardened ground in the Panxi River basin of 20%, 35%, 20% and 25%, respectively, and with the combination of the water quality characteristics of runoff from various underlying surfaces, the values of pollutant concentrations were as shown in Table 6.10.

The source runoff non-point source pollutants calculated based on the pollution load as shown in Table 6.10, collection area of each water body and ground surface property were as shown in Table 6.11.

Figure 6.9 Distribution of sewage outfalls in Jiangbei District.

4. Total exogenous pollution in the basin.

   Based on the data of point source pollution (Table 6.8) and non-point source pollution (Table 6.11) in the basin, the total exogenous pollution of each water body in the Panxi River basin was as shown in Table 6.12.

6.2.1.2 Investigation on endogenous pollution sources

1. Investigation on existing situation of endogenous load.

   Due to perennial input of masses of pollutants into lakes, reservoirs and rivers in the Panxi River basin, slow flow velocity of water and long water change cycles of

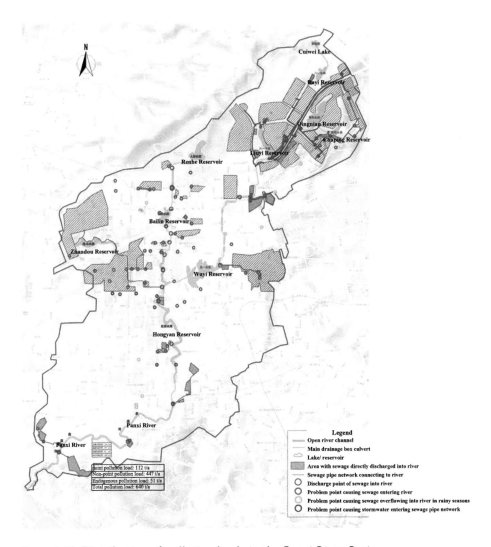

Figure 6.10 Distribution of pollution loads in the Panxi River Basin.

most lakes and reservoirs, lots of silt deposited on the bottoms of rivers, lakes and reservoirs, resulting in endogenous pollution.

According to field measurement results, the silt area, depth and volume of each lake, reservoir and river were as shown in Table 6.13.

2. Endogenous pollution load.

According to relevant data, the average sediment pollutant release intensity $(kg/m^3/a)$ of lakes or rivers in Chongqing was as shown in Table 6.14.

The sediment pollutant release load of the Panxi River and each lake and reservoir in the basin calculated based on the above coefficients was as shown in Table 6.15.

Table 6.6 Sewage volume received by each water body in the Panxi River Basin

Water body	Sewage receiving range (km²)	Served population (10⁴)	Max. daily sewage volume (10⁴ m³/d)	Average daily sewage volume (10⁴ m³/d)	Annual sewage volume (10⁴ m³)
Cuiwei Lake	0.05	0.08	0.02	0.02	7.3
Bayi Reservoir	0.00	0.00	0.00	0.00	0
Qingnian Reservoir	0.00	0.00	0.00	0.00	0
Chaping Reservoir	0.00	0.00	0.00	0.00	0
Liuyi Reservoir	0.00	0.00	0.00	0.00	0
Renhe Reservoir	0.00	0.00	0.00	0.00	0
Bailin Reservoir	0.00	0.00	0.00	0.00	0
Zhandou Reservoir	0.10	0.15	0.06	0.05	18.25
Wuyi Reservoir	2.46	3.69	1.49	1.24	452.6
Hongyan Reservoir	0.63	0.95	0.33	0.28	102.2
Panxi River	1.19	1.79	0.66	0.55	200.75
Total	4.44	6.66	2.56	2.13	777.45

Table 6.7 Average concentrations of pollutants in Chongqing domestic sewage

Item	COD	NH3-N	TN	TP
Average concentration (mg/L)	320	30	40	3.5

Table 6.8 Point source pollution load of each water body

Water body	COD (t/a)	NH_3-N (t/a)	TN (t/a)	TP (t/a)
Cuiwei Lake	23.4	2.2	2.9	0.3
Bayi Reservoir	0.0	0.0	0.0	0.0
Qingnian Reservoir	0.0	0.0	0.0	0.0
Chaping Reservoir	0.0	0.0	0.0	0.0
Liuyi Reservoir	0.0	0.0	0.0	0.0
Renhe Reservoir	0.0	0.0	0.0	0.0
Bailin Reservoir	0.0	0.0	0.0	0.0
Zhandou Reservoir	58.4	5.5	7.3	0.6
Wuyi Reservoir	1,448.3	135.8	181.0	15.8
Hongyan Reservoir	327.0	30.7	40.9	3.6
Panxi River	642.4	60.2	80.3	7.0
Total	2,487.8	233.2	311.0	27.2

Note: The overflow of pipe networks in incompletely separate systems was unable to be estimated, so the table only showed the domestic sewage volume at the pollution discharge point in the dry season and did not include the pollution load caused by mixed stormwater and sewage in incompletely separate systems in the rainy season.

Table 6.9 Pollutant indicators for stormwater runoff in Chongqing

Type of underlying surface	CODcr (mg/L)		TSS (mg/L)		TN (mg/L)		TP (mg/L)		NH$_3$-N (mg/L)	
	Average value	Initial runoff	Average value	Initial runoff	Average value	Initial runoff	Average value	Initial runoff	Average value	Initial runoff
Natural rainfall	/	23	/	100	/	3.0	/	0.03	/	1.2
Concrete pavement	90	100–250	280	300–1,000	6.9	6–12	0.56	0.6–1.2	1.0	3–5
Asphalt pavement	120	200–350	560	1,000–1,500	3.87	5–10	0.71	0.7–1.2	1.3	3–6
Asphalt roof	74.3	100–150	200	150–250	3.6	5–10	0.16	0.3–0.4	2.7	3–5
Tiled roof	48	50–100	37	60–150	4.0	6–10	0.12	0.2–0.3	1.2	4–6
Cement roof	77	180–450	65	65–395	5.6	15–31	0.2	3–3.5	1.8	7–11
Community road	40	120–200	400	1,600–2,000	3.8	6–8	0.28	0.4–0.6	2.0	3–5
Concrete plaza in business district	60	100–150	500	1,000–1,500	3.2	5–8	0.3	1–2	2.5	5–7
Green space	60.5	97.8	22.4	650	2.85	3.2	0.44	0.21	1.18	1.6

Table 6.10 Calculation parameters for surface runoff pollution load in the Panxi
River Basin

Item	Typical pollutant type	Average concentration in stormwater runoff (mg/L)	Rainfall (m)	Runoff coefficient
Pavement	CODcr	120	1.08	0.9
	TN	3.87	1.08	0.9
	TP	0.71	1.08	0.9
	NH$_3$-N	1.3	1.08	0.9
Green space	CODcr	60.5	1.08	0.2
	TN	2.85	1.08	0.2
	TP	0.44	1.08	0.2
	NH$_3$-N	1.18	1.08	0.2
Roof	CODcr	66.4	1.08	0.9
	TN	4.4	1.08	0.9
	TP	0.16	1.08	0.9
	NH$_3$-N	1.9	1.08	0.9
Hardened ground	CODcr	60	1.08	0.9
	TN	3.2	1.08	0.9
	TP	0.3	1.08	0.9
	NH$_3$-N	2.5	1.08	0.9

Table 6.11 Total pollutant discharge of urban surface runoff in the Panxi River Basin

Water body	Collection area (km^2)	COD (t/a)	NH$_3$-N (t/a)	TN (t/a)	TP (t/a)
Cuiwei Lake	1.00	26.04	0.33	0.73	0.12
Bayi Reservoir	0.30	20.54	0.48	1.00	0.12
Qingnian Reservoir	0.60	36.53	0.93	1.16	0.13
Chaping Reservoir	0.40	27.65	0.65	1.34	0.16
Liuyi Reservoir	2.09	144.43	3.41	7.01	0.82
Renhe Reservoir	0.45	14.90	0.35	0.78	0.12
Bailin Reservoir	1.28	44.60	2.03	4.11	0.47
Zhandou Reservoir	0.10	6.93	0.16	0.27	0.04
Wuyi Reservoir	5.30	366.08	8.64	17.77	2.07
Hongyan Reservoir	6.14	227.53	8.13	14.13	1.65
Panxi River	6.90	476.87	11.25	23.15	2.70
Total	24.54	1,392.09	36.36	71.43	8.40

6.2.2 Analysis on characteristics of pollution sources in the Panxi River basin

According to the filed investigation, survey and estimation of pollution sources, the pollutants in the Panxi River basin mainly came from the water (including domestic sewage and surface runoff) from box culverts, surface runoff around the lakes and river and sediment release. Pollution sources and endogenous loads in the Panxi River basin were summarized in Table 6.16.

According to the calculation results, among the pollution source loads of Panxi River, the domestic sewage point source pollution load was the highest, the stormwater runoff also accounted for a larger proportion, while the endogenous release was relatively low.

Table 6.12 Exogenous pollution load in the Panxi River Basin

Water body	Indicator (t/a)	Exogenous point source	Exogenous non-point source	Total
Cuiwei Lake	COD	23.4	26.0	49.4
	NH$_3$-N	2.2	0.3	2.5
	TN	2.9	0.7	3.6
	TP	0.3	0.1	0.4
Bayi Reservoir	COD	0	20.5	20.5
	NH$_3$-N	0	0.5	0.5
	TN	0	1.0	1
	TP	0	0.1	0.1
Qingnian Reservoir	COD	0	36.5	36.5
	NH$_3$-N	0	0.9	0.9
	TN	0	1.2	1.2
	TP	0	0.1	0.1
Chaping Reservoir	COD	0	27.6	27.6
	NH$_3$-N	0	0.7	0.7
	TN	0	1.3	1.3
	TP	0	0.2	0.2
Liuyi Reservoir	COD	0	144.4	144.4
	NH$_3$-N	0	3.4	3.4
	TN	0	7.0	7
	TP	0	0.8	0.8
Renhe Reservoir	COD	0	14.9	14.9
	NH$_3$-N	0	0.3	0.3
	TN	0	0.8	0.8
	TP	0	0.1	0.1
Bailin Reservoir	COD	0	44.6	44.6
	NH$_3$-N	0	2.0	2
	TN	0	4.1	4.1
	TP	0	0.5	0.5
Zhandou Reservoir	COD	58.4	6.9	65.3
	NH$_3$-N	5.5	0.2	5.7
	TN	7.3	0.3	7.6
	TP	0.6	0.0	0.6
Wuyi Reservoir	COD	1,448.3	366.1	1,814.4
	NH$_3$-N	135.8	8.6	144.4
	TN	181	17.8	198.8
	TP	15.8	2.1	17.9
Hongyan Reservoir	COD	327	227.5	554.5
	NH$_3$-N	30.7	8.1	38.8
	TN	40.9	14.1	55
	TP	3.6	1.6	5.2
Panxi River	COD	642.4	476.9	1,119.3
	NH$_3$-N	60.2	11.3	71.5
	TN	80.3	23.1	103.4
	TP	7	2.7	9.7
Total in the basin	COD	2,487.8	1,392.09	3,879.89
	NH$_3$-N	233.2	36.36	269.56
	TN	311	71.43	382.43
	TP	27.2	8.40	35.6

Table 6.13 Results of investigation on sediment in the Panxi River Basin

Water body	Silt area (10^4 m^2)	Average silt depth (m)	Silt volume (10^4 m^3)
Cuiwei Lake	1.00	0.50	0.50
Bayi Reservoir	2.08	0.25	0.52
Qingnian Reservoir	6.88	0.12	0.82
Chaping Reservoir	1.88	0.20	0.38
Liuyi Reservoir	5.87	0.15	0.90
Renhe Reservoir	5.54	0.10	0.55
Bailin Reservoir	3.13	0.08	0.25
Zhandou Reservoir	2.32	0.25	0.60
Wuyi Reservoir	9.3	0.26	2.40
Hongyan Reservoir	6.3	0.50	3.15
Panxi River	10.25	0.30	3.08
Total	63.57		13.15

Table 6.14 Average sediment pollutant release intensity in the Panxi River Basin

Item	COD	NH_3-N	TN	TP
Lake/reservoir	0.32	0.0018	0.0036	0.0007
River	1	0.0036	0.0072	0.0014

Table 6.15 Sediment pollutant release load of each water body in the Panxi River Basin

Water body	COD (t/a)	NH_3-N (t/a)	TN (t/a)	TP (t/a)
Cuiwei Lake	1.60	0.01	0.018	0.004
Bayi Reservoir	1.66	0.01	0.019	0.004
Qingnian Reservoir	2.62	0.01	0.030	0.006
Chaping Reservoir	1.22	0.01	0.014	0.003
Liuyi Reservoir	2.88	0.02	0.032	0.006
Renhe Reservoir	1.76	0.01	0.020	0.004
Bailin Reservoir	0.80	0.00	0.009	0.002
Zhandou Reservoir	1.92	0.01	0.022	0.004
Wuyi Reservoir	7.68	0.04	0.086	0.017
Hongyan Reservoir	10.08	0.06	0.113	0.022
Panxi River	30.80	0.11	0.222	0.043
Total	63.02	0.29	0.58	0.11

6.2.3 Water environment problems in the Panxi River basin and the cause analysis

According to the above investigation and analysis on the pollution in the Panxi River basin, the water environment in the Panxi River basin had the following problems.

6.2.3.1 Existing problems of exogenous pollution and the cause analysis

1. Imperfect pipe network construction, incomplete separation of stormwater and sewage and direct discharge of sewage.
   As the construction of municipal drainage facilities in the Panxi River basin lagged behind the urban development, some secondary and tertiary pipe networks in the box culvert service area realized the separation of stormwater and sewage, but the mixing of stormwater and sewage still existed in some secondary and tertiary pipe networks. Meanwhile, a large amount of mixed stormwater and sewage overflowed from pollutant intercepting main pipe at the end of drainage pipe. Besides, there was also the phenomenon that water from sewage pipe network was directly discharged into the river, with the preliminary estimated directly discharged sewage volume of about 27,000 m^3, most of which entered and accumulated in lakes and reservoirs in the basin for a long term. On the whole, the point source pollution was serious in the basin, which was a major cause of the black and odorous water in the Panxi River basin.
2. Discharge of a large amount of construction wastewater.
   There were many construction sites in the Panxi River basin. As a result, a large amount of construction wastewater was directly discharged into Panxi River without sedimentation treatment, seriously affecting the water body appearance.
3. Small interception ratio of main pipe and large number of annual overflow times.
   The incompletely separate systems were adopted before Hongyan Reservoir and after Wuyi Reservoir. The interception ratios of combined pollutant intercepting main pipes were low, almost all below 1.5. Therefore, there were frequent overflows in rainy seasons and large amounts of mixed stormwater and sewage entered the river, lakes and reservoirs, causing pollution.

   Serious pollution of urban surface runoff, pollutants entering lakes without purification.

The area within the Panxi River basin was basically built up, with a high hardening rate, where the stormwater runoff pollution was serious. The initial stormwater around lakes and rivers without any treatment directly entered into the river with a lot of pollutants carried through surface runoff, seriously affecting the water quality.

6.2.3.2 Existing problems of endogenous pollution and the cause analysis

The water pollution was aggravated due to the lack of endogenous pollution treatment. A large amount of stormwater and domestic sewage entered water bodies in the Panxi River basin, making silt easily settling at the bottom. Some lakes and reservoirs had been dredged, but the dredging was of little significance for the pollution load reduction if the domestic sewage was not completely intercepted. Therefore, in addition

Table 6.16 Summary of pollution source loads in the Panxi River Basin

Water body	Indicator (t/a)	Exogenous point source	%	Exogenous non-point source	%	Endogenous	%	Total
Cuiwei Lake	COD	23.4	45.88	26	50.98	1.600	3.14	51.000
	NH$_3$-N	2.2	87.68	0.3	11.96	0.009	0.36	2.509
	TN	2.9	80.15	0.7	19.35	0.018	0.50	3.618
	TP	0.3	74.35	0.1	24.78	0.004	0.87	0.404
Bayi Reservoir	COD	0	0.00	20.5	92.49	1.664	7.51	22.164
	NH$_3$-N	0	0.00	0.5	98.16	0.009	1.84	0.509
	TN	0	0.00	1	98.16	0.019	1.84	1.019
	TP	0	0.00	0.1	96.49	0.004	3.51	0.104
Qingnian Reservoir	COD	0	0.00	36.5	93.29	2.624	6.71	39.124
	NH$_3$-N	0	0.00	0.9	98.39	0.015	1.61	0.915
	TN	0	0.00	1.2	97.60	0.030	2.40	1.230
	TP	0	0.00	0.1	94.57	0.006	5.43	0.106
Chaping Reservoir	COD	0	0.00	27.7	95.79	1.216	4.21	28.916
	NH$_3$-N	0	0.00	0.7	99.03	0.007	0.97	0.707
	TN	0	0.00	1.3	98.96	0.014	1.04	1.314
	TP	0	0.00	0.2	98.69	0.003	1.31	0.203
Liuyi Reservoir	COD	0	0.00	144.4	98.04	2.880	1.96	147.280
	NH$_3$-N	0	0.00	3.4	99.53	0.016	0.47	3.416
	TN	0	0.00	7	99.54	0.032	0.46	7.032
	TP	0	0.00	0.8	99.22	0.006	0.78	0.806
Renhe Reservoir	COD	0	0.00	14.9	89.44	1.760	10.56	16.660
	NH$_3$-N	0	0.00	0.4	97.58	0.010	2.42	0.410
	TN	0	0.00	0.8	97.58	0.020	2.42	0.820
	TP	0	0.00	0.1	96.29	0.004	3.71	0.104

(Continued)

Table 6.16 (Continued) Summary of pollution source loads in the Panxi River Basin

Water body	Indicator (t/a)	Exogenous point source	%	Exogenous non-point source	%	Endogenous	%	Total
Bailin Reservoir	COD	0	0.00	44.6	98.24	0.800	1.76	45.400
	NH$_3$-N	0	0.00	2	99.78	0.005	0.22	2.005
	TN	0	0.00	4.1	99.78	0.009	0.22	4.109
	TP	0	0.00	0.5	99.65	0.002	0.35	0.502
Zhandou Reservoir	COD	58.4	86.88	6.9	10.26	1.920	2.86	67.220
	NH$_3$-N	5.5	96.31	0.2	3.50	0.011	0.19	5.71
	TN	7.3	95.78	0.3	3.94	0.022	0.28	7.622
	TP	0.6	99.30	0	0.00	0.004	0.70	0.604
Wuyi Reservoir	COD	1,448.3	79.49	366.1	20.09	7.680	0.42	1,822.080
	NH$_3$-N	135.8	94.02	8.6	5.95	0.043	0.03	144.443
	TN	181	91.01	17.8	8.95	0.086	0.04	198.886
	TP	15.8	88.19	2.1	11.72	0.017	0.09	17.917
Hongyan Reservoir	COD	327	57.92	227.5	40.30	10.080	1.79	564.580
	NH$_3$-N	30.7	79.01	8.1	20.85	0.057	0.15	38.857
	TN	40.9	74.21	14.1	25.58	0.113	0.21	55.113
	TP	3.6	67.64	1.7	31.94	0.022	0.41	5.322
Panxi River	COD	642.4	55.86	476.9	41.47	30.800	2.68	1,150.100
	NH$_3$-N	60.2	84.07	11.3	15.78	0.111	0.15	71.611
	TN	80.3	77.42	23.2	22.37	0.222	0.21	103.722
	TP	7	71.85	2.7	27.71	0.043	0.44	9.743
Total in the basin	COD	2,487.8	63.10	1,392.1	35.31	63.024	1.60	3,942.924
	NH$_3$-N	233.2	86.40	36.4	13.49	0.292	0.11	269.892
	TN	311	81.20	71.4	18.64	0.584	0.15	382.984
	TP	27.2	76.16	8.4	23.52	0.114	0.32	35.714

to exogenous pollution, there were also endogenous pollution problems. Besides, the water bodies were formed for many years, during which the sediments caused by sewage input and algae growth accumulated at the river bottom and formed polluted sediments, aggregating the deterioration trend of water quality.

6.2.3.3 Existing problems of environmental management mechanism and the cause analysis

The coordination mechanism was not smooth due to the lack of a basin collaboration system. Exogenous inputs of the Panxi River basin involved large areas such as Huangshan Avenue, Crystal Central, Evergrande Palace, Lake Paradise, Jinkai Avenue, Palm Springs Community, Hongmian Avenue, Jinxiu Shanzhuang, Jinlong Road, Songqiao Road, Donghe Chuntian, Heji Primary School and Panxi Road, namely, the whole Panxi River basin stretches over Liangjiang New Area, Yubei District and Jiangbei District, but no unified basin management mechanism was established, resulting in a lack of collaboration among districts.

6.3 Practice of regional water environment treatment[3]

6.3.1 Treatment idea

The Panxi River basin covers a large area of 28.25 km^2, involving many lakes and reservoirs, so the treatment scheme should be planned from the perspective of the whole basin. The Panxi River was faced with two problems: pollution and lack of clean water supplement. According to the field investigation and analysis, the pollution of Panxi River mainly came from four sources, which, in the order of pollution severity, respectively, were: direct sewage discharge from surrounding urban living quarters, overflow pollution of incompletely separate systems, initial stormwater non-point source pollution and endogenous pollution of lakes and reservoirs. Therefore, the source control and pollutant interception and basin water supplement were key projects for the Panxi River treatment. Besides, strong support for the healthy river water from ecosystems around and within the lakes and reservoirs was needed, realizing the sound interaction among plants, humans and water bodies. From the perspective of functions, first, the building of green peripheral ecological barriers could effectively reduce the surface runoff's direct impacts on river water; second, it could also further realize the closeness between human and nature and increase the benign impact of human activities on water bodies, which is also the practical significance of lake and river treatment.

Therefore, various engineering measures in the two implementation schemes developed this time would closely focus on the above three aspects, by following the treatment principles that: the effectiveness of exogenous pollutant interception is a prerequisite for the success of river treatment; the core way to solve the lake and river pollution is to restore the self-purification function of water bodies and shorten the water change cycle; and the technology strategy of "taking ecological and non-ecological measures as the primary and supplementary measures respectively" shall be adopted.

The principles of design for treatment projects included manure and reliable technologies, convenient maintenance and management, low operating cost, integration with landscape and livable ecology.

According to the administrative division and pollution situation in the basin and in order to facilitate the treatment work, the Panxi River basin was divided into four control units in the scheme, as shown in Figure 6.11.

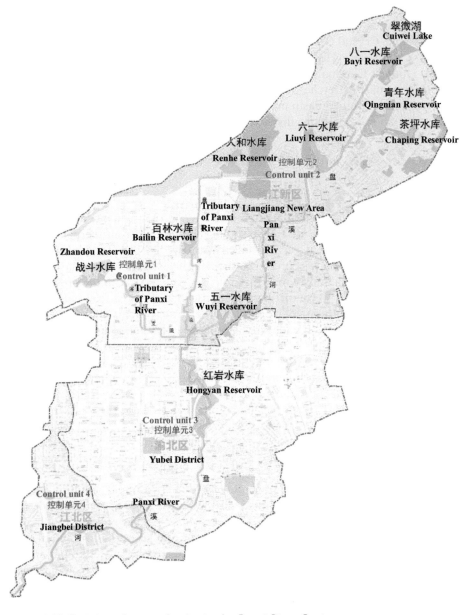

Figure 6.11 Division of control units in the Panxi River Basin.

6.3.2 Environmental capacity and reduction target

6.3.2.1 Environmental capacity

The water environmental capacity refers to the maximum allowable quantity of pollutants in a water body per unit time under a given water scope and hydrological conditions and on the premises of specified pollution drainage way and water quality goals, reflecting the water's capacity for receiving pollutants on the premise of sustainable and normal functions of the basin water environment system. The basin water environment carrying capacity is generally determined by an appropriate water environment model based on the water quality goal, certain hydrological and hydrodynamic conditions, spatial distribution of pollution discharge, etc.

The water environmental capacity of the Panxi River was calculated by a one-dimensional model, with the formula as shown below:

$$W = 31.536 \times \alpha \times \left(C_S \cdot e^{Kx/(86.4 \times u)} - C_R \right) Q_R \qquad (6.1)$$

Wherein: W—environmental capacity, t/a

C_R—water quality concentration of water from the upstream, mg/L

α—coefficient of non-uniformity, depending on the river width, the value taken here is 1

C_S—concentration along the river, mg/L

Q_R—flow rate of water from the upstream, m^3/s

u—average flow velocity at the river cross section, m/s

K—degradation coefficient, 1/d

x—distance along the river, m (Table 6.17).

Major water bodies in the Panxi River basin were mainly for landscape and flood control, so the planned water quality goal was Class V. The reservoir environmental capacity calculated based on the abovementioned model and planned water quality goal was as shown in Table 6.18.

6.3.2.2 Reduction target

Based on the calculated results of water environmental capacity and estimated pollutants input, the total reduction quantity of major pollutions in river could be obtained.

There was no remaining environmental capacity in most of the lakes, reservoirs and river segments in the Panxi River basin, and according to the water quality goal

Table 6.17 Estimation of Panxi River water quality model parameters

Parameter	Parameter meaning	Estimated value	Unit
KC	COD integrated attenuation coefficient	0.2	1/d
KN	NH$_3$-N integrated attenuation coefficient	0.1	1/d
KTN	TN integrated attenuation coefficient	0.2	1/d
KTP	TP integrated attenuation coefficient	0.05	1/d

Table 6.18 Calculated results of environmental capacity

		COD	NH₃-N	TN	TP
Planned water quality goal (mg/L)		40	2	2	0.2
Environmental capacity (t/a)	Cuiwei Lake	32.0	0.4	0.6	0.1
	Bayi Reservoir	37.1	0.5	0.5	0.1
	Qingnian Reservoir	145.6	1.0	1.1	0.2
	Chaping Reservoir	136.2	2.1	1.5	0.1
	Liuyi Reservoir	51.3	0.6	0.6	0.1
	Renhe Reservoir	156.6	2.6	2.4	0.2
	Bailin Reservoir	108.2	0.6	0.6	0.1
	Zhandou Reservoir	165.3	2.7	2.7	0.2
	Wuyi Reservoir	663.6	15.3	10.1	1.0
	Hongyan Reservoir	131.4	4.6	7.1	0.7
	Panxi River	250.2	8.2	10.8	1.1
	Total	1,877.6	38.6	38.1	3.8

of Class V, the quantities of major pollutants COD, NH₃-N, TN and TP to be reduced were as listed in Table 6.19. The volume of domestic sewage overflow into water bodies in rainy seasons due to incompletely separate systems was unable to be measured during the calculation of pollution load. Therefore, the pollution load reduction rates listed in the table were calculated on the basis that there was no overflow in rainy seasons after the overflow outlet renovation.

6.3.3 Treatment scheme

The Scheme I for integrated treatment of the Panxi River basin focused on the "source control and pollutant interception", with the main engineering systems including: source control and pollutant interception, ecological water supplement, lake/ reservoir/ river water quality improvement, river ecological restoration and reconstruction and water quality monitoring. The system diagram of Scheme I was as shown in Figure 6.12:

6.3.3.1 Source control and pollutant interception work

The source control and pollutant interception work in the scheme involved the reconstruction of 23.5 km pipe network in the basin, characterized by wide range, long construction line, complex construction conditions and large work quantity. Relevant local department should gradually complete the separation of stormwater and sewage and sewage outfall renovation before the end of 2017, then the self-purification function of water bodies could be gradually restored.

1. Pipe network treatment.
   It was newly found that about 10.48 km sewage pipe network (including 5,456 m in Liangjiang New Area, 4,762 m in Yubei District and 258 m in Jiangbei District) and 13.01 km stormwater pipe network (including 7,706 m in Liangjiang New Area, 5,132 m in Yubei District and 237 m in Jiangbei District) needed reconstruction.

Table 6.19 Reduction of major pollutants in the Panxi River Basin

Water body	Indicator (t/a)	Pollutant load (t/a)				Environmental capacity (t/a)	Annual reduced load (t/a)	
		Exogenous point source	Exogenous non-point source	Endogenous	Total		Reduction quantity	Reduction rate (%)
Cuiwei Lake	COD	23.4	26	1.600	51.00	32	19	37.25
	NH$_3$-N	2.2	0.3	0.009	2.51	0.4	2.109	84.06
	TN	2.9	0.7	0.018	3.62	0.6	3.018	83.42
	TP	0.3	0.1	0.004	0.40	0.1	0.3035	75.22
Bayi Reservoir	COD	0	20.5	1.664	22.16	16.5	5.664	25.55
	NH$_3$-N	0	0.5	0.009	0.51	0.5	0.00936	1.84
	TN	0		0.019	1.02	0.5	0.51872	50.92
	TP	0	0.1	0.004	0.10	0.05	0.05364	51.76
Qingnian Reservoir	COD	0	36.5	2.624	39.12	62	/	/
	NH$_3$-N	0	0.9	0.015	0.91	1	/	/
	TN	0	1.2	0.030	1.23	1.1	0.12952	10.53
	TP	0	0.1	0.006	0.11	0.2	/	/
Chaping Reservoir	COD	0	27.7	1.216	28.92	22	6.916	23.92
	NH$_3$-N	0	0.7	0.007	0.71	0.6	0.10684	15.12
	TN	0	1.3	0.014	1.31	0.6	0.71368	54.33
	TP	0	0.2	0.003	0.20	0.1	0.10266	50.66
Liuyi Reservoir	COD	0	144.4	2.880	147.28	156.6	/	/
	NH$_3$-N	0	3.4	0.016	3.42	2.6	0.8162	23.89
	TN	0	7	0.032	7.03	2.4	4.6324	65.87
	TP	0	0.8	0.006	0.81	0.2	0.6063	75.20
Renhe Reservoir	COD	0	14.9	1.760	16.66	43.9	/	/
	NH$_3$-N	0	0.4	0.010	0.41	0.6	0.2198	26.81
	TN	0	0.8	0.020	0.82	0.6	0.00385	3.71
	TP	0	0.1	0.004	0.10	0.1	/	/

(Continued)

Table 6.19 (Continued) Reduction of major pollutants in the Panxi River Basin

Water body	Indicator (t/a)	Pollutant load (t/a)				Environmental capacity (t/a)	Annual reduced load (t/a)	
		Exogenous point source	Exogenous non-point source	Endogenous	Total		Reduction quantity	Reduction rate (%)
Bailin Reservoir	COD	0	44.6	0.800	45.40	68.3	/	/
	NH₃-N	0	2	0.005	2.00	2.1	/	/
	TN	0	4.1	0.009	4.11	1.5	2.609	63.49
	TP	0	0.5	0.002	0.50	0.1	0.40175	80.07
Zhandou Reservoir	COD	58.4	6.9	1.920	67.22	69.8	/	/
	NH₃-N	5.5	0.2	0.011	5.71	2.7	3.0108	52.72
	TN	7.3	0.3	0.022	7.62	2.7	4.9216	64.57
	TP	0.6	0	0.004	0.60	0.2	0.4042	66.90
Wuyi Reservoir	COD	1,448.3	366.1	7.680	1,822.08	663.6	1,158.48	63.58
	NH₃-N	135.8	8.6	0.043	144.44	15.3	129.1432	89.41
	TN	181	17.8	0.086	198.89	10.1	188.7864	94.92
	TP	15.8	2.1	0.017	17.92		16.9168	94.42
Hongyan Reservoir	COD	327	227.5	10.080	564.58	131.4	433.18	76.73
	NH₃-N	30.7	8.1	0.057	38.86	4.6	34.2567	88.16
	TN	40.9	14.1	0.113	55.11	7.1	48.0134	87.12
	TP	3.6	1.7	0.022	5.32	0.7	4.62205	86.85
Panxi River	COD	642.4	476.9	30.800	1,150.10	250.2	899.9	78.25
	NH₃-N	60.2	11.3	0.111	71.61	8.2	63.41088	88.55
	TN	80.3	23.2	0.222	103.72	10.8	92.92176	89.59
	TP	7	2.7	0.043	9.74	1.1	8.64312	88.71
Total in the basin	COD	2,487.8	1,392.1	63.024	3,942.92	1516.50	2,426.324	61.54
	NH₃-N	233.2	36.3	0.292	269.89	38.6	231.29214	85.70
	TN	311	71.4	0.584	382.98	38.1	344.88428	90.05
	TP	27.2	8.4	0.114	35.71	3.7	32.01361	89.64

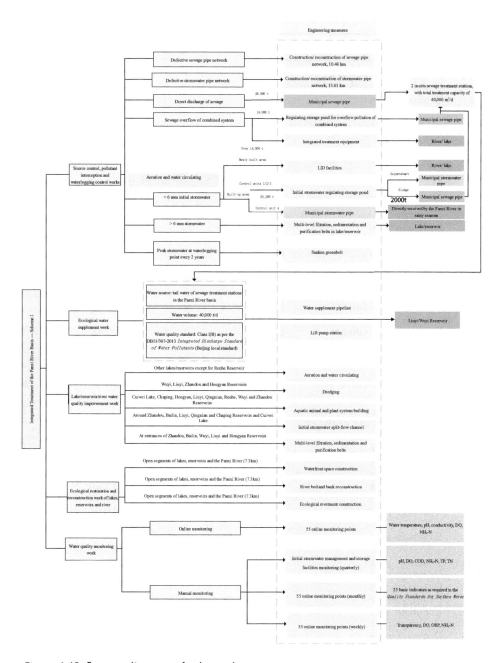

Figure 6.12 System diagram of scheme I.

After the separate stormwater and sewage system reconstruction, the sewage interception rate in the Panxi River basin should be considered at 80% in view of the incomplete separation of stormwater and sewage in some municipal and community facilities (Table 6.20).

Table 6.20 Engineering quantity list of pipe network treatment in the Panxi River Basin

S/N	Engineering name	Engineering content	Unit	Quantity
(I)	Sewage pipe network treatment work			
1	Drainage pipe	DN1000, fiberglass reinforced plastic mortar pipe	m	771
2	Drainage pipe	DN300, HDPE double-wall corrugated pipe	m	721
3	Drainage pipe	DN350, HDPE double-wall corrugated pipe	m	94
4	Drainage pipe	DN400, HDPE double-wall corrugated pipe	m	4,979
5	Drainage pipe	DN500, HDPE double-wall corrugated pipe	m	1,317
6	Drainage pipe	DN600, HDPE double-wall corrugated pipe	m	935
7	Drainage pipe	DN700, HDPE double-wall corrugated pipe	m	48
8	Drainage pipe	DN800, fiberglass reinforced plastic mortar pipe	m	886
9	Drainage box culvert	L × B = 1,000 × 1,000 mm, reinforced concrete	m	8
10	Drainage box culvert	L × B = 1,200 × 1,500 mm, reinforced concrete	m	30
11	Drainage box culvert	L × B = 1,200 × 1,200 mm, reinforced concrete	m	15
12	Drainage box culvert	L × B = 800 × 600 mm, reinforced concrete	m	672
13	Drainage inspection well	D1000, reinforced concrete	No.	336
(II)	Stormwater pipe network treatment work			
1	Drainage pipe	DN400, double-wall corrugated pipe	m	148
2	Drainage pipe	DN500, double-wall corrugated pipe	m	991
3	Drainage pipe	DN600, double-wall corrugated pipe	m	2,083
4	Drainage pipe	DN300, double-wall corrugated pipe	m	5
5	Drainage pipe	DN700, double-wall corrugated pipe	m	349
6	Drainage pipe	DN800, fiberglass reinforced plastic mortar pipe	m	3,868
7	Drainage pipe	DN1000, fiberglass reinforced plastic mortar pipe	m	2,903
8	Drainage pipe	DN1200, reinforced concrete	m	450
9	Drainage pipe	DN1400, reinforced concrete	m	685
10	Drainage pipe	DN1500, reinforced concrete	m	657
11	Drainage pipe	DN1600, reinforced concrete	m	256
12	Drainage pipe	DN2000, reinforced concrete	m	324
13	Drainage pipe	DN2400, reinforced concrete	m	143
14	Drainage pipe	DN2200, reinforced concrete	m	94
15	Drainage pipe	B × H = 1,000 × 500, reinforced concrete	m	8
16	Drainage pipe	B × H = 1,200 × 800, reinforced concrete	m	113
17	Drainage pipe	DN1000, reinforced concrete	No.	443

2. Overflow pollution control work.

In order to control the non-point source pollution, reduce peak flow in drainage pipes and avoid puddles formed on ground, the scheme planned to newly build 11 regulating storage ponds for the pollution control of incompletely separate pipe network systems within the scope, with the total reservoir capacity of 14,000 m³. In view of the actual situation, the regulating storage ponds for the overflow pollution control of incompletely separate systems were adopted with the mode of centralized processing and regulation to regulate the sewage in rainy seasons through main pipes. The regulating storage ponds were mainly for the overflow pollution control of incompletely separate systems, with the effective volume calculation formula as below:

$$V = 3600 \, t_1 (n - n_0) Q_{dr} \beta \tag{6.2}$$

Wherein: V—effective volume of the regulating storage pond (m³)

t_1—inflow time of regulating storage pond (h), preferably in the range of 0.5–1.0 hour: the upper limit should be adopted if the quality of overflow sewage in an incompletely separate drainage system during a single rainfall has no obvious initial effect; otherwise the lower limit should be adopted.

n—interception ratio of the regulating storage pond after being built and put into service, obtained based on the required pollution load reduction target, local interception ratio and the proportion of interception volume in rainfall

n_0—original interception ratio of the system

Q_{dr}—previous dry weather sewage flow of intercepting well (m³/s)

β—safety coefficient, in the range of 1.1–1.5.

There were 11 regulating storage ponds for overflow pollution control of incompletely separate systems in the Panxi River basin, including four ones in Liangjiang New Area, with the capacity of 5,091 m³; five ones in Yubei District, with the capacity of 7,341 m³; and two ones in Jiangbei District, with the capacity of 1,503 m³; with the total capacity of 13,935 m³ (Figure 6.13; Table 6.21).

Figure 6.13 Real picture of overflow pollution of incompletely separate systems in the Panxi River Basin.

Table 6.21 Engineering quantity list of regulating storage ponds for pollution control of incompletely separate systems in the Panxi River Basin

S/N	Administrative district	Regulating storage pond for overflow pollution control	Volume (m³)	Area (m²)
1	Yubei District	WL7	50	13
2	Yubei District	W9-1	1,220	139
3	Yubei District	W9-2	2,089	239
4	Yubei District	W10	2,762	316
5	Yubei District	W11	1,220	195
6	Jiangbei District	W5	136	36
7	Jiangbei District	W6	1,367	219
8	Liangjiang New Area	W1	2,005	229
9	Liangjiang New Area	W2	1,309	150
10	Liangjiang New Area	W3	394	79
11	Liangjiang New Area	W8	1,383	158
Total			13,935	1,773

3. Water purification stations.

Water purification stations were used to treat domestic sewage that was directly discharged into Panxi River before and mixed sewage collected by regulating storage ponds for overflow pollution control. Two water purification stations were designed in the scheme, namely, Tianhu Park water purification station (15,000 t/d) and Dongbu Park water purification station (25,000 t/d), with the total capacity of 40,000 t/d.

① Tianhu Park water purification station

a. Design scale

The Tianhu Park water purification station was designed to collect the domestic sewage around Liuyi Reservoir and from the upstream, with the design scale Q = 15,000 m³/d.

b. Site selection and form

The selected site was located within Tianhu Park, adjacent to Jinshan Avenue on the west. It was a fully-buried underground water purification station.

c. Process flow

The process flow was as shown in Figure 6.14:

d. Scheme design

The designed sewage treatment capacity of the water purification station was 15,000 t/d on sunny days with no variation coefficient considered and was 19,500 t/d in rainy seasons with a variation coefficient of 1.3 considered. The integrated sewage treatment room was a fully-buried underground structure, with the underground building covering an area of 4,674 m². The management room was set up above the ground, covering an area of 368 m². The influent and effluent quality of the water purification station was as shown in Table 6.22:

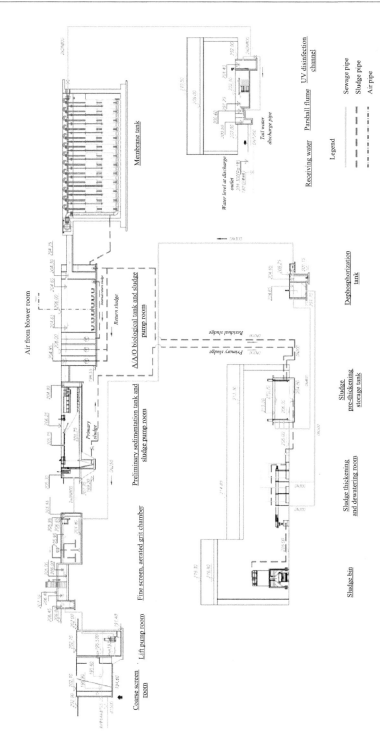

Figure 6.14 Treatment process flowchart of Tianhu Park sewage station.

Table 6.22 Influent and effluent quality of Tianhu Park water
purification station

S/N	Item name	Influent quality	Effluent quality	Unit
1	PH	6–9	6–9	
2	COD	<430	30	mg/L
3	BOD	<170	6	mg/L
4	SS	<230	5	mg/L
5	TN	<45	15	mg/L
6	NH₃-N	<35	1.5 (2.5)	mg/L
7	TP	<6	0.3	mg/L

The Tianhu Park sewage station was adopted with the treatment process of AAO biochemical pool + membrane bioreactor (MBR), meeting the treatment effect as required in the design, and the process flow was as shown in Figure 6.15:

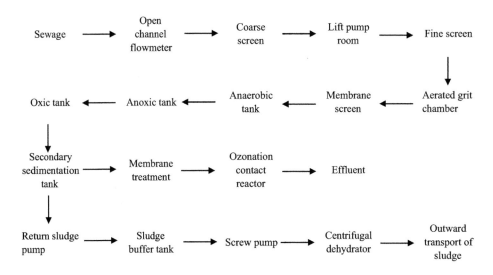

Figure 6.15 Process flowchart of Tianhu Park sewage treatment station.

The sewage intercepted from DN1200 sewage main pipe is transported to the coarse screen via the municipal road sewage pipe network after flowing through the open channel flowmeter; enters the fine screen by being lifted via lift pump; enters the AAO biochemical pool after passing the aerated grit chamber and membrane screen; passes through the horizontal secondary sedimentation tank and membrane treatment room, then the ozonation contact reactor and finally is lifted to the park water system or used to supplement the river system. The sludge produced will enter the sludge buffer tank first, then will be lifted by screw pump to the centrifugal dehydrator after being cut by sludge shredder and the mud cakes reaching a certain moisture content will be transported by sludge transport trunk.

② Dongbu Park water purification station

a. Design scale
Dongbu Park water purification station was designed to collect the domestic sewage around Wuyi Reservoir and from the upstream, with the design scale $Q = 25,000$ m^3/d.

b. Site selection and form
The selected site was located within Dongbu Park, close to the side of Xingguang Avenue. It was a fully-buried underground water purification station.

c. Process flow
The process flow was as shown in Figures 6.16 and 6.17.

d.   Scheme design
The designed sewage treatment capacity of the sewage treatment plant was 25,000 t/d on sunny days with no variation coefficient considered, and was 32,500 t/d in rainy seasons with a variation coefficient of 1.3 considered. The integrated sewage treatment room was a fully-buried underground structure, with the underground building covering an area of 9,268 m^2. The management room was set up above the ground, covering an area about 3,200 m^2. The influent and effluent quality of the water purification station was as shown in Table 6.23.

The Dongbu Park sewage station was adopted with the treatment process of AAO biochemical pool + MBR, meeting the treatment effect as required in the design, and the process flow was as shown in Figure 6.18.

The sewage intercepted from the DN1200 sewage main pipe firstly flows to the coarse screen after flowing through the open channel flowmeter; enters the fine screen by being lifted via lift pump; enters the AAO biochemical pool after passing the aerated grit chamber and membrane screen; passes through the horizontal secondary sedimentation tank and membrane treatment room, then the ozonation contact reactor and finally is lifted to the park water system or used to supplement the river system. The sludge produced will enter the sludge buffer tank first, then will be lifted by screw pump to the centrifugal dehydrator after being cut by sludge shredder and the mud cakes reaching a certain moisture content will be transported by sludge transport trunk.

6.3.3.2 Initial stormwater control work[8][12]

1. Initial stormwater management and storage in the basin.
The non-point source pollution means that dissolved and solid pollutants gather in receiving water bodies from non-specific sites through the runoff process under the effects of runoff or snowmelt scouring. With the improvement of point source pollution control rate, the urban non-point source pollution's contribution to the water pollution gradually emerges, which has become an important source of water pollution. The US EPA (1993) has listed urban surface runoff as the third largest pollution source of rivers and lakes across the United States and also reported nationwide that urban runoff was a major source of the surface water quality

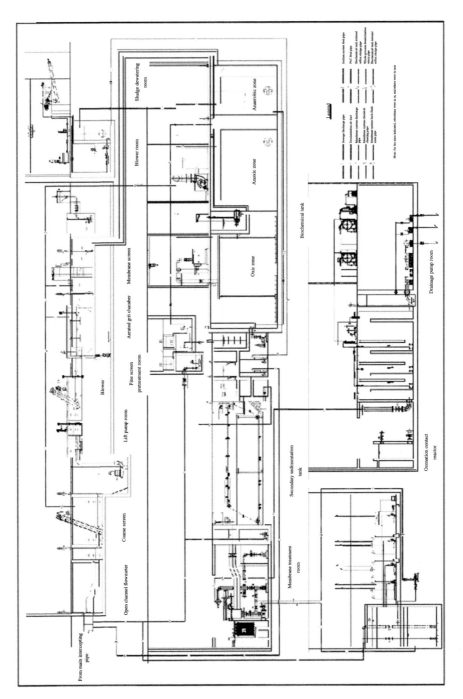

Figure 6.16 Treatment process flowchart of Dongbu Park sewage station.

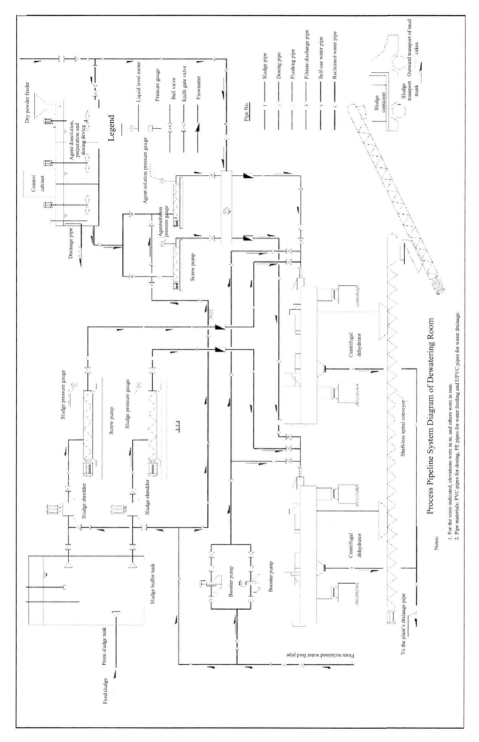

Figure 6.17 Sludge process flowchart of Dongbu Park.

Table 6.23 Influent and effluent quality of Dongbu Park water purification station

S/N	Item name	Influent quality	Effluent quality	Unit
1	PH	6–9	6–9	
2	COD	<430	30	mg/L
3	BOD	<170	6	mg/L
4	SS	<230	5	mg/L
5	TN	<45	15	mg/L
6	NH₃-N	<35	1.5 (2.5)	mg/L
7	TP	<6	0.3	mg/L

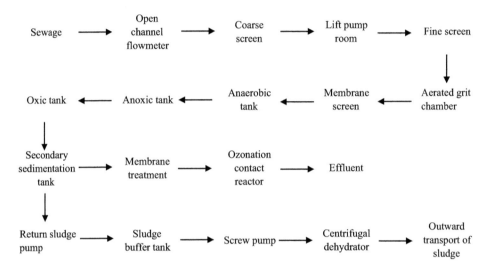

Figure 6.18 Process flowchart of Dongbu Park sewage treatment station.

deterioration (US EPA, 2000), and among the analysis on eight pollution sources, the contribution rates of urban runoff to the water quality deterioration of rivers, lakes and estuaries ranked the sixth, fourth and second, respectively. Comparatively, the study on non-point source pollution started relatively late in China that it was not until the 1970s that the study on urban storm runoff pollution began. Due to the prominent contradiction of point source pollution in China, the research on non-point source pollution developed slowly. After 2000, with the vigorous promotion of urbanization process in China and large increase in point source pollution control rate, the urban storm runoff pollution gradually received extensive attention of domestic scholars. Especially, the national major special project for water pollution control set up from 2006 to 2010 greatly promoted the research process of urban non-point source pollution in China, and the research on urban non-point source pollution was successively carried out by many cities, such as Zhengzhou, Shanghai, Wuhan, Guangzhou and Xiamen. All results of China's investigation and research on the polluted areas of major rivers and lakes such as the Dianchi Lake, Taihu Lake and Huaihe River basins indicate that non-point source pollution has become an important factor of water pollution, among which the results of the research on the eutrophication of the Dianchi Lake show that the contribution rates of industrial wastewater, urban sewage and non-point source pollution are 9%, 24% and 67%, respectively.

① Initial stormwater collection scale

At present, there is no relatively unified and accurate method to calculate the initial rainfall volume in China, and the following is the initial stormwater collection scale of some cities in China:

As stipulated in Article 5.6.3 of the *Engineering Technical Code for Rain Utilization in Building and Sub-district* (GB50400-2006): the initial runoff split-flow shall be determined based on the concentrations of pollutants such as the COD, SS and chromaticity of stormwater collected on the underlying surface; where no data are available, the runoff thickness ranges of 2–3 and 3–5 mm can be adopted, respectively, for the roof and surface runoff split-flow.

Beijing: Usually during the same rain, the initial rainfall on road surface is more than that on roofs. When the net rainfall of initial roof stormwater is about 2–3 mm, more than 60% of the total stormwater runoff pollution load during the rain can be controlled, while the effect is not obviously improved when the controlled net rainfall is above 3 mm. The net rainfall of initial stormwater on road surfaces varies greatly, but the runoff pollution is relatively slight when the runoff net rainfall is 7–8 mm.

Handan: The initial roof and road surface runoff split-flow depths determined based on the stormwater pollutant wash-off model building and establishment of the exponential relationship between pollutant concentration and rainfall by sampling analysis are 3 mm and 6 mm, respectively, which can, respectively, remove 78.3% and 77.9% of the total COD.

Zhengdong New District: According to the empirical estimation, the initial rainfall runoff depth is considered at 3 mm.

Shenzhen: In case of no measured runoff and pollutant load concentration curve data available and by reference to the *Engineering Technical Code for Rain Utilization in Building and Sub-district*, the roof runoff interception depth will be 3–5 mm, and the surface runoff interception depth will be 5–7 mm.

Japan: As for the overflow pollution control in areas with incompletely separate drainage systems, the regulation and storage volume per unit area is 2–7 mm.

Requirements in Chongqing in 2014: As for the problem of initial stormwater collection, initial stormwater regulating storage ponds should be set up to collect the surface stormwater with a depth of 4–6 mm and covering an area of 1–2 km^2, which should be mainly set up at the connection to river water system at the downstream end of stormwater pipeline.

By reference to relevant interception standards of cities in China and combining with the urban environment, climate, city structure and social and economic development of Chongqing, the rainfall interception depth is considered at 6 mm or at 4 mm for stormwater zone with locally restricted land.

② Schemes for initial stormwater management and storage facilities

   a. Set up initial stormwater split-flow wells and construct new intercepting pipes

As indicated in Figure 6.19, an initial stormwater split-flow well is set up in the stormwater branch pipe line connected to the main drainage box culvert. The instantaneous peak flow rate in the intercepting pipe during the split-flow well's collection of 6 mm water in the collection area range can be calculated by the rainstorm intensity formula, then the maximum hourly design flow rate can be calculated based on the sewage pipe connected to the intercepting pipe. Based on the aforesaid two flow rates, it can be estimated if the current sewage pipe network can bear the maximum hourly design flow rate and the initial stormwater with a depth of 6 mm at the same time under the most unfavorable conditions. If the current sewage pipe network capacity is insufficient, initial stormwater intercepting pipes shall be constructed separately.

   b. Set up initial stormwater regulating storage ponds

The initial stormwater regulating storage pond is set up in the stormwater branch pipe line connected to the main drainage box culvert, with the initial stormwater with a depth of 6 mm collected within the scope of the stormwater branch pipe's collection area as the pond volume. The stormwater is subject to flocculation sedimentation treatment in the pond, then the supernatant will be discharged into the stormwater pipe network, while the flocculent sludge will be discharged into the municipal sewage pipe by the submersible sewage pump in the pond. The brief scheme steps are shown in Figure 6.20.

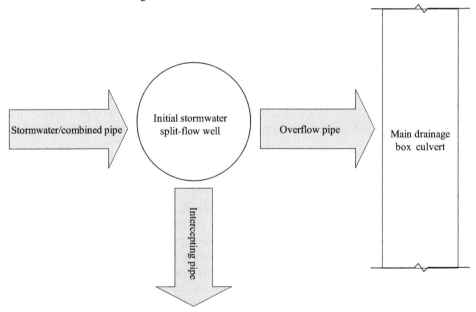

Figure 6.19 Schematic diagram of initial stormwater split-flow well.

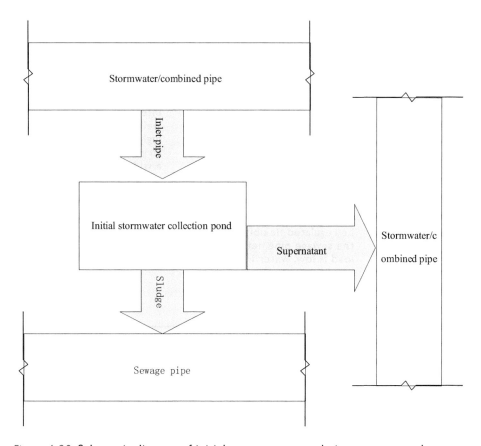

Figure 6.20 Schematic diagram of initial stormwater regulating storage pond.

    c.   Scheme comparison (Table 6.24)

        According to the design idea of Scheme I, it shall be ensured that no initial stormwater enters the lakes and reservoirs, but the Panxi River downstream of Hongyan Reservoir can receive initial stormwater. Therefore, the initial stormwater shall be controlled in Control Unit 1, Control Unit 2 and area upstream of Hongyan Reservoir dam in Control Unit 3, and initial stormwater regulating storage ponds of high safety level shall be adopted to control the initial stormwater.

③   Initial stormwater pollution load

    The research on the initial stormwater quality started earlier in foreign countries, and some scholars put forward the proportions of initial stormwater runoff pollutants in rainfall, as shown in Table 6.25.

    There also have been some research results in China in recent years. For example, the maximum loads of TSS, COD, TN and TP in the initial 30% runoff in Shenyang were 75%, 79%, 90% and 60%, respectively (Li Chunlin et al.); the initial 40% runoff in Chongqing Huxi basin carried 60%–80% of the pollution load (Wang Shumin et al.); the initial 20%–30% road surface runoff

Table 6.24 Comparison of schemes for initial stormwater management and storage facilities

Facility	Advantages	Disadvantages
Initial stormwater split-flow well + initial stormwater intercepting pipe	The intercepting wells in built-up area occupy small space, have low requirements for underground space and can save the investment. The intercepting pipes can be constructed along rivers, where relatively broad land is available.	During the process of collecting 6 mm initial stormwater by the intercepting pipe in split-flow well, the peak flow in the pipe is high, which requires a relatively large-diameter pipe and involves a wide construction scope.
Initial stormwater regulating storage pond	The discharge of initial stormwater in the pond can be regulated flexibly. When the sewage pipe network load is low, water in the pond can be pumped into the sewage pipe network by booster pump, which is of relatively high safety.	It covers a relatively large area, has high requirements for underground space and requires higher investment.

Table 6.25 Proportions of initial stormwater runoff pollutants in rainfall

Item	Proportion of initial runoff volume P_1/%	Total proportion of pollutants in initial runoff P_2/%	Time of being put forward
Stahre and Urbonas	20	>80	1990
Wanielista and Yousef	25	>50	1993
Hsaget et al.	30	>80	1995
Bertrand-Krajewski et al.	30	>80	1998
Deletic	20	>40	1998

in Guangzhou carried 50%–60% of the total COD, SS and heavy metals (Gan Huayang et al.); the initial 40% road surface runoff carried 53%±16% TSS, 66%±10% COD, 59%±2% TN and 51%±5% TP and the initial 40% runoff carried 64%±20% TSS, 66%±17% COD, 55%±14% TN and 56%±3% TP (He Qiang et al.).

④ Design for initial stormwater regulating storage pond
The calculation of initial stormwater volume is as below:

$$V = 10\Psi Ah \tag{6.3}$$

V—initial stormwater volume (m^3)
Ψ—integrated runoff coefficient

A—service area of the stormwater system (hm^2)

h—intercepted rainfall (mm)

Different runoff standards are adopted for different places. According to the planned building density, four zone properties can be divided. The integrated runoff coefficients of zones of the four properties are as shown in Table 6.26.

According to the estimation, the initial stormwater volume in the water collection area upstream of Wuyi Reservoir in the Panxi River basin was 27,202 m^3.

According to the stormwater collection paths, the area upstream of Hongyan Reservoir dam in the Panxi River basin was divided into 27 small stormwater management zones and an initial stormwater regulating storage pond was set up at the drainage end of each zone, namely 27 initial stormwater regulating storage ponds were set up dispersedly, with the detailed volume, covering area and diameter of the upstream stormwater pipe of each pond as shown in Table 6.27. Based on the site implementation conditions, the water collection area was 1,056 ha, the total volume was 63,200 m^3 and the covering area was 14,856 m^2.

The specific distribution of the regulating storage ponds was as shown in Figure 6.21.

According to the above domestic and foreign studies on initial stormwater pollution load, the stormwater pollution load reduction rate was considered at 60% after the initial stormwater interception.

6.3.3.3 Ecological water supplement work

The segment pollutant interception, segment treatment and segment water supplement were adopted in the ecological work supplement work to utilize reclaimed water for the ecological water supplement on the basis of source control and pollutant interception, so as to construct the ecological water supplement system of the whole basin. The core of ecological water supplement work lay in the layout of water purification stations, where the surrounding sewage was treated in dry seasons and then directly discharged into lakes and reservoirs for ecological water supplement. The scheme utilized the tail water of Tianhu Park and Dongbu Park water purification stations for the water supplement of Liuyi Reservoir and Wuyi Reservoir respectively, with the water supplement volumes of 15,000, 25,000 t/d, respectively. The discharge standard of tail water met the Class I(B) standard as stipulated in the *Integrated Discharge Standard of Water Pollutants of Beijing* (DB11/307-2013). The ancillary facilities of the water supplement work were as shown in Table 6.28.

Table 6.26 Integrated runoff coefficient

S/N	Zone property (area of impermeable covering layer, %)	Integrated runoff coefficient
I	Central zone with dense buildings (70%)	0.6–0.8
2	Residential zone with relatively dense buildings (50%–70%)	0.5–0.7
3	Residential zone with relatively sparse buildings (30%–50%)	0.4–0.6
4	Residential zone with very sparse buildings (30%)	0.3–0.5

Table 6.27 Parameters of initial stormwater regulating storage pond in each collection area

S/N	Regulating storage pond no.	Local administrative district	Local basin	Collection area (10⁴ m²)	Pond volume (m³)	Effective depth (m)	Flat area of pond (m²)
1	A1	Liangjiang New Area	Wuyi Reservoir	64.87	3,892	5	778
2	A2	Liangjiang New Area	Wuyi Reservoir	23.76	1,425	5	285
3	A3	Liangjiang New Area	Wuyi Reservoir	65.32	3,919	5	784
4	A4	Liangjiang New Area	Wuyi Reservoir	25.52	1,531	5	306
5	A5	Liangjiang New Area	Wuyi Reservoir	15.15	909	5	182
6	A6	Liangjiang New Area	Wuyi Reservoir	49.42	2,965	5	593
7	A7	Liangjiang New Area	Wuyi Reservoir	47.43	2,846	5	569
8	A9	Liangjiang New Area	Liuyi Reservoir	47.10	2,826	5	565
9	A10	Liangjiang New Area	Liuyi Reservoir	55.34	3,320	5	664
10	A13	Liangjiang New Area	Chaping Reservoir	27.46	1,647	5	329
11	A14	Liangjiang New Area	Liuyi Reservoir	17.66	1,059	5	212
12	A15	Liangjiang New Area	Wuyi Reservoir	17.20	860	5	172
13	A16	Liangjiang New Area	Bailin Reservoir	20.52	1,231	3	410
14	A17	Liangjiang New Area	Bailin Reservoir	94.74	5,684	5	1,137
15	A19	Yubei District	Hongyan Reservoir	32.68	1,961	3	654
16	A20	Yubei District	Hongyan Reservoir	30.35	1,821	3	607
17	A21	Yubei District	Hongyan Reservoir	20.51	1,231	3	410
18	A22	Yubei District	Hongyan Reservoir	16.87	1,012	3	337
19	A23	Yubei District	Hongyan Reservoir	38.08	2,285	4	571
20	A24	Yubei District	Hongyan Reservoir	20.14	1,208	3	403
21	A25	Liangjiang New Area	Hongyan Reservoir	12.56	754	3	251
22	A26	Liangjiang New Area	Hongyan Reservoir	81.18	4,871	5	974
23	A27	Liangjiang New Area	Hongyan Reservoir	29.69	1,781	3	594
24	A28	Liangjiang New Area	Hongyan Reservoir	26.75	1,605	3	535
25	A29	Liangjiang New Area	Hongyan Reservoir	78.69	4,721	5	944
26	A30	Liangjiang New Area	Hongyan Reservoir	25.81	1,549	3	516
27	A31	Liangjiang New Area	Hongyan Reservoir	71.40	4,284	4	1,071
	Total				63,197		14,853

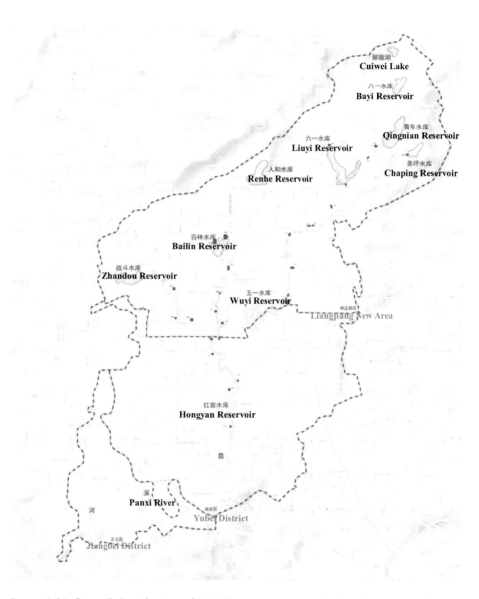

Figure 6.21 Overall distribution of initial stormwater regulating storage ponds.

6.3.3.4 Lake and reservoir water quality improvement work

1. Improvement goals.

   Clear water (transparency >1.2 m), normal color, no unpleasant odor and full display of water landscape
   Rich in aquatic species and abundant bio-diversity
   Healthy, stable and long-lasting operating water ecosystem.

Table 6.28 Ancillary facilities of ecological water supplement engineering

S/N	Engineering content	Specification/scale	Unit	Quantity
I	I# reclaimed water supplement pump station	15,000 m³/d	No.	I
2	2# reclaimed water supplement pump station	25,000 m³/d	No.	I
3	Reclaimed water supplement pipe	DN300, steel pipe	m	1,273
4	Reclaimed water supplement pipe	DN450, steel pipe	m	1,950

Table 6.29 Water quality improvement scheme of lakes and rivers in the Panxi River Basin

Water body	Water quality level	Water quality improvement scheme
Bayi Reservoir Qingnian Reservoir Chaping Reservoir Renhe Reservoir	Level I	Split-flow channel + micro-ecological in situ purification technology
Cuiwei Lake Bailin Reservoir	Level 2	Split-flow channel + micro-ecological in situ purification technology
Zhandou Reservoir Wuyi Reservoir Hongyan Reservoir Liuyi Reservoir	Level 3	Split-flow channel + dredging + micro-ecological in situ purification technology + engineering bacteria adding

2. Improvement schemes.

A water quality improvement system for the Panxi River basin was established that focused on ecological restoration and was supplemented by endogenous reduction and stormwater purification. According to different water quality conditions of each lake, reservoir and the Panxi River, the ten lakes and reservoirs and the Panxi River were classified into three treatment levels and different pertinent water quality improvement schemes were put forward, as indicated in Table 6.29:

① Treatment of Level-1 lakes and reservoirs

The Level-1 lakes and reservoirs included the Renhe Reservoir, Bayi Reservoir, Qingnian Reservoir and Chaping Reservoir, primarily located at the sources of each sub-basin, with relatively good water quality currently and relatively single pollution source (point source pollution had been basically treated and the non-point source pollution was the primary source) and not affected by the upstream pollution. The treatment idea was to enhance the self-purification capacity of water bodies and the treatment measures included split-flow channel and micro-ecological in-situ purification technology (Figure 6.22).

The direct purification process of micro-ecology in situ purification technology uses the biological contact oxidation mechanism to treat landscape water by the three core technologies, namely, water circulating, oxygenic aeration ecological and microbial purification. The direct purification process core of the micro in situ purification

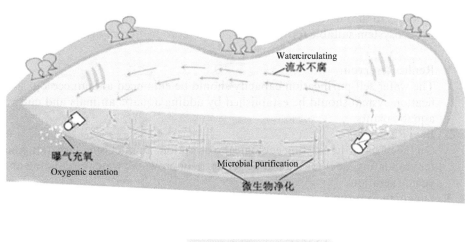

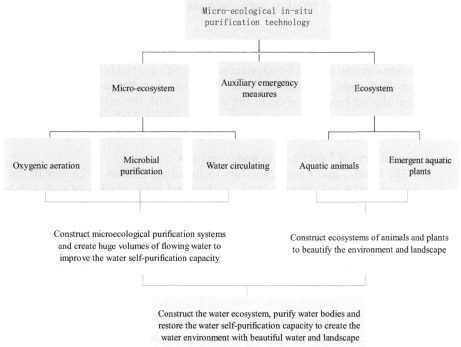

Figure 6.22 Schematic diagram micro-ecological in situ purification technology.

technology is the ecological water control, and the process flows focus on creating favorable survival environment and suitable water purifying environment for water-purifying microorganisms. The micro-ecological in situ purification technology utilizes the vital activities of microorganisms and takes the natural microorganisms as the core to improve the activity of microorganisms by in situ breeding and combing with technological means of water circulating and oxygenic aeration to transfer, transform and degrade pollutants in water and restore the self-purification capacity of water bodies to the maximum extent, so as to purify the water, rebuild and restore the aquatic ecosystem suitable for the survival and reproduction of diverse species.

1. Renhe Reservoir.
   The water self-purification capacity should be enhanced and an ecological purification system should be established by adding aquatic animals and cultivating aquatic plants.
2. Bayi Reservoir.
   On the basis of the existing ecological restoration, an integrated aeration and water circulating machine should be set up to improve the water circulating and oxygenic aeration.
3. Qingnian Reservoir.
   The micro-ecological in situ purification system should be established in the reservoir, with the main measures including integrated aeration and water circulating machine, biological blanket, artificial aquatic mat, aquatic animal and aquatic plant.
4. Chaping Reservoir.
   By making use of existing surrounding stormwater pipe networks, a split-flow infiltration channel should be constructed around the reservoir to purify the stormwater discharged into the reservoir. The micro-ecological in situ purification system should be established in the reservoir, with the main measures including integrated aeration and water circulating machine, biological blanket, artificial aquatic mat, aquatic animal and aquatic plant.

   ② Treatment of Level-2 lakes and reservoirs
   The Level-2 lakes and reservoirs included the Cuiwei Lake and Bailin Reservoir, primarily located in the middle and lower reaches of each sub-basin, with point source pollution and non-point source pollution as the main pollution sources. The pollution sources were hard to be completely eliminated, but the pollution load was relatively low. The treatment idea was to enhance the self-purification capacity of water bodies and the treatment measures included split-flow channel and micro-ecological in situ purification technology.

1. Cuiwei Lake.
   By making use of existing surrounding stormwater pipe networks, a split-flow infiltration channel should be constructed around the lake to purify the stormwater discharged into the lake. The micro-ecological in situ purification system should be established in the reservoir, with the main measures including integrated aeration and water circulating machine, biological blanket, artificial aquatic mat, aquatic animal and aquatic plant.
2. Bailin Reservoir.
   By making use of existing surrounding stormwater pipe networks, a split-flow infiltration channel should be constructed around the lake to purify the stormwater discharged into the reservoir. On the basis of the existing ecological restoration, an integrated aeration and water circulating machine should be set up to improve the water circulating and oxygenic aeration.

③ Treatment of Level-3 lakes and reservoirs
   The Level-3 lakes and reservoirs included the Zhandou Reservoir, Wuyi Reservoir and Liuyi Reservoir, primarily located in the lower reaches of each sub-basin, with complex pollution situation and affected by the sewage input from the upstream. The main pollution sources included point source pollution and non-point source pollution, which were hard to be treated. The treatment idea was to restore the self-purification capacity of water bodies and the treatment measures included split-flow channel + dredging + micro-ecological in situ purification technology + engineering bacteria adding.

1. Dredging.
   Due to the perennial input of mixed stormwater and sewage from the upstream, the reservoirs were deposited with thick silt layers and the samples collected were black and odorous, so it was necessary to clean up the reservoirs.
   There are many methods for the reservoir silt treatment, which can be classified into physical treatment, biological treatment, biochemical treatment and silt solidification. The physical treatment method includes measures such as dredging, washing and natural restoration, with the advantages of quick effects, low technical requirements and simple construction procedures, and the disadvantages of large impacts on and interference with the public, and secondary pollution easily caused by transportation and dumping. The biological treatment method mainly utilizes the degradation ability of microorganisms or bacteria to degrade the pollutants in silt or utilizes the plants' capabilities of gathering, absorbing and degrading organic pollutants to purify the silt. Although it requires low investment, it has slow effect and is easily affected by external environmental conditions. The biochemical method is to inject chemicals into silt to let the chemicals react with organics and other pollutants in the silt, so as to change the chemical properties of pollutants and improve the microorganisms' activity and capability of degrading organics. It is characterized by quick and steady effects, but it can't thoroughly remove pollutants from water for water bodies with difficulties in changing water and limited environmental capacity. As for the silt solidification method, the silt is taken as the raw material, and additives such as soil coagulant are added to mix with the silt by mechanical means to change structural properties of the silt, then silt will be solidified, dehydrated and reduced in a short time. This method is characterized by short treatment period, simple process, volume reduction of silt after solidification, relatively steady physical properties and no secondary pollution caused during the transportation and dumping.
   In addition to the above classification, the methods can also be classified into in situ treatment and ex situ treatment according to the treatment sites. The surface of in situ sampling results was black and odorous sludge deposited from sewage over the years, giving off a foul smell. According to the actual site investigation, the reservoirs should be dredged, with the mechanical dredging as the specific improvement method and outward transportation as the sludge disposal method.
   According to the site investigation results, the dredging volumes of each reservoir were as follows: Zhandou Reservoir, 23,000 m^3; Wuyi Reservoir, 189,000 m^3; Liuyi Reservoir, 59,000 m^3 and Hongyan Reservoir, 63,000 m^3.

3. Micro-ecological in situ purification technology.
   The micro-ecological in situ purification technology includes the following aspects, namely, aeration and water circulating, biological blanket, artificial aquatic mat, aquatic animal and aquatic plant.

Aeration and water circulating: in order to facilitate the up-and-down water flowing in the reservoir to prevent pollutants accumulating in bottom under weak water flowing circumstance and accelerate the downward transfer efficiency of surface reaeration, an integrated aeration and water circulating machine with an extra-large flow is set up in the deepest area in front of the dam, which can provide dissolved oxygen for the water body and also promote the water circulation to (1) transfer the dissolved oxygen to everywhere and avoid anoxic zones, (2) make the water fully contact with the biological membrane for better mass transfer and (3) inhibit algae blooms.

Biological blanket and artificial aquatic mat: the biological blanket and artificial aquatic mat are made of polymer composite materials with the structures imitating blanket and branches and leaves of aquatic weeds, which can float freely in the water, forming the vertical structure of upper, middle and lower layers. The biological blanket and artificial aquatic mat are characterized by porous structures and high specific surface areas, with microorganisms gathering on their surfaces, forming the micro-environment with the "oxic–facultative–anaerobic" composite structure and realizing the nitration and denitrification.

Aquatic animal and aquatic plant: they can build the aquatic ecosystem. The aquatic animal community building is to further restore the species diversity by building aquatic animal communities and promote the micro-circulation in water by adding selected and cultured aquatic animals, such as fish, shrimp, snail and shellfish, so as to create more favorable conditions for the growth of other aquatic life. Aquatic plant communities can provide food sources and habitats for hydrophilous aquatic birds, insects and other wildlife. Precisely because of the interaction and circulation between aquatic animals and plants and abiotic substances, the water body becomes an aquatic ecological environment with the life vitality, thus preserving the biodiversity in the aquatic environment, which is the basis for the performance of other functions. The aquatic plants can absorb carbon dioxide in the environment and release oxygen by photosynthesis, and at the same time, they can also absorb many harmful elements in water, thereby eliminating pollution, purifying water, improving the water quality and restoring ecological functions of water. For example, water hyacinths can absorb the nitrogen, phosphorus, potassium and heavy metal ions, while reeds not only can purify the suspended solids, chloride, organic nitrogen and sulfate in water but also can absorb the mercury and lead in water.

Zhandou Reservoir: 23 integrated aeration and water circulating machines, 5,750 m^2 biological blankets, 218,000 m artificial aquatic mats, 60,000 kg aquatic animals and 2,300 m^2 aquatic plants.

Wuyi Reservoir: 69 integrated aeration and water circulating machines, 31,500 m^2 biological blankets, 888,000 m artificial aquatic mats, 100,000 kg aquatic animals and 12,600 m^2 aquatic plants.

Liuyi Reservoir: 30 integrated aeration and water circulating machines, 14,750 m^2 biological blankets, 209,000 m artificial aquatic mats, 80,000 kg aquatic animals and 5,900 m^2 aquatic plants.

Hongyan Reservoir: 30 integrated aeration and water circulating machines, $14,000 \, m^2$ biological blankets, $104,000 \, m$ artificial aquatic mats, $70,000 \, kg$ aquatic animals and $7,000 \, m^2$ aquatic plants (Table 6.30).

6.3.3.5 River ecological restoration and reconstruction work

As for the ecological restoration and reconstruction of "three-side bare" (hardened river bottom and both banks) open river segments (with the total length of 7.3 km) of the Panxi River, the river channel ecological functions of the Panxi River should be restored by means of river bed excavation + plant covering + bank surface reconstruction + gravel bed to improve the water self-purification capacity. Meanwhile, the landscape on both river banks should be designed by adding facilities such as waterfront footpath and landscape stones to provide leisure and entertainment places for citizens to enjoy the results of environment treatment.

6.3.3.6 Water quality monitoring work

For the supervision over treatment results, the black and odorous water monitoring points should be established in the basin. In principle, a detection point should be set every 200–600 m along the black and odorous water body, but no less than three detection points should be set for each water body. The sampling point was generally set at 0.5 m below the water surface, and set at 1/2 of the water depth in case of the water depth less than 0.5 m. In principle, the detection should be conducted once every 2–8 days, for at least three times in total. The water quality monitoring work in the scheme included automatic monitoring and manual detection. A total of 55 online monitoring points were set, including 36 in Liangjiang New Area, 11 in Yubei District and 8 in Jiangbei District. The manual detection was mainly to monitor the initial stormwater management and storage facilities. The water quality routine detection should be conducted quarterly (on six indicators, including pH, DO, COD, NH_3-N, TP and TN). The manual sampling monitoring on 25 basic indicators as required in the *Quality Standards for Surface Water* should be conducted monthly for the 55 online monitoring points. Conventional judgment indicators for black and odorous water should be monitored weekly, with the indicators including transparency, DO, oxidation–reduction potential (ORP) and NH_3-N. The arrangement of monitoring points was as shown in Figure 6.23, work quantities in Table 6.31 and the general layout for integrated treatment in Figure 6.24.

6.4 Evaluation on regional water environment effects

The evaluation on water environment treatment helps the effective realization of treatment goals. The water environment treatment plays a vital role in the urban development. In order to promote sustainable urban development, we shall use scientific methods to evaluate the operation of urban water environment economic system and follow certain scientific principles while evaluating the water environment treatment. The evaluation on water environment can effectively reflect the operation situation of urban environmental system and change trends of economy, resources and environment, so as to gradually improve aspects unfavorable to water environment, let the

Table 6.30 Water quality improvement engineering quantities for lakes and reservoirs in the Panxi River Basin

S/N	Engineering name	Engineering content	Quantity	Unit
1	Cuiwei Lake			
1.1	Split-flow infiltration channel around the lake		471	m
1.2	Integrated aeration and water circulating machine		6	Set
1.3	Biological blanket		2,000	m^2
1.4	Artificial aquatic mat		25,000	m
1.5	Aquatic animals		15,000	kg
1.6	Aquatic plant		1,000	m^2
2	Bayi Reservoir			
2.1	Integrated aeration and water circulating machine		6	Set
3	Qingnian Reservoir			
3.2	Integrated aeration and water circulating machine		20	Set
3.3	Biological blanket		17,250	m^2
3.4	Artificial aquatic mat		191,000	m
3.5	Aquatic animals		70,000	kg
3.6	Aquatic plant		6,900	m^2
4	Chaping Reservoir			
4.1	Split-flow infiltration channel around the lake		695	m
4.2	Integrated aeration and water circulating machine		10	Set
4.3	Biological blanket		4,750	m^2
4.4	Artificial aquatic mat		67,000	m
4.5	Aquatic animals		15,000	kg
4.6	Aquatic plant		1,900	m^2
5	Liuyi Reservoir			
5.1	Sediment dredging		0.8	$10^4\ m^3$
5.2	Compound engineering bacteria		41.8	t
5.3	Split-flow infiltration channel around the lake		2,501	m
5.4	Integrated aeration and water circulating machine		30	Set
5.5	Biological blanket		14,750	m^2
5.6	Artificial aquatic mat		209,000	m
5.7	Aquatic animals		80,000	kg
5.8	Aquatic plant		5,900	m^2
6	Renhe Reservoir			
6.1	Aquatic animals		80,000	kg
6.2	Aquatic plant		5,500	m^2

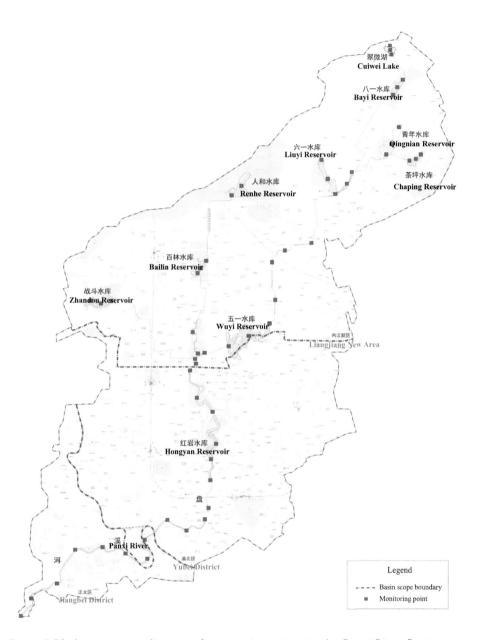

Figure 6.23 Arrangement diagram of monitoring points in the Panxi River Basin.

whole environment system return to the benign development track, improve the water environment treatment efficiency and promote all-round urban development.

The evaluation on water environment treatment results is not only an assessment or a phased summary but also can promote the treatment, rectification and maintenance, which is of great significance for the water environment treatment.

Table 6.31 Engineering quantity list of automatic monitoring system for the Panxi River Basin

Lake/reservoir name	Unit	Automatic monitoring device quantity	Engineering investment (10^4 yuan)
Bayi Reservoir	Set	3	60
Bailin Reservoir	Set	3	60
Chaping Reservoir	Set	3	60
Cuiwei Lake	Set	3	60
Hongyan Reservoir	Set	5	100
Liuyi Reservoir	Set	5	100
Panxi River	Set	22	440
Qingnian Reservoir	Set	3	60
Renhe Reservoir	Set	2	40
Wuyi Reservoir	Set	3	60
Zhandou Reservoir	Set	3	60
Integrated information system	Set	3	3,000
Total		58	4,100

The promotion of treatment by evaluation can accelerate treatment progresses. The third-party institutions can know about the treatment progress and results of each river in the evaluation scope by investigation and pre-evaluation prior to evaluation, then analyze the treatment progress of each river according to national requirements for river treatment progress in different cities. As for any treatment project with slow progress or not achieving results, a notice of early warning shall be circulated to urge to accelerate the treatment progress.

The promotion of rectification by evaluation can improve treatment results. According to pre-evaluation results, the third-party institutions shall analyze the treatment results of each river on the principles of "adaptation to local conditions and one strategy for one river" and based on their experience in water environment treatment and protection. As for any treatment project not reaching the expected results, opinions on the adjustment and improvement shall be put forward to effectively promote them to achieve the treatment goals.

The promotion of maintenance by evaluation can maintain treatment effects. Based on pre-evaluation results, the third-party institutions shall put forward suggestions for improvement and maintenance of river treatment effects for rivers achieving "preliminary effects" but with unstable treatment effects or room for improvement according to relevant national requirements for the "long-term clear water" stage, so as to make plans for the realization of "long-term clear water" goal in advance.

The Panxi River originates from Liangjiang New Area of Chongqing, flowing through Liangjiang New Area, Yubei District and Jiangbei District. The Panxi River basin covers an area about 28.13 km^2, and there are ten reservoirs such as the Cuiwei Lake and Bayi Reservoir in the upstream. The river water is mainly from the abovementioned reservoirs and stormwater in the basin. Before treatment, due to reasons such as combined stormwater and sewage systems of the upstream drainage box culverts, stormwater runoff pollution in the box culvert service area, incomplete elimination of endogenous pollution of the Panxi River, the water quality of the Panxi River gradually deteriorated and the basin was full of sludge. There were six river segments

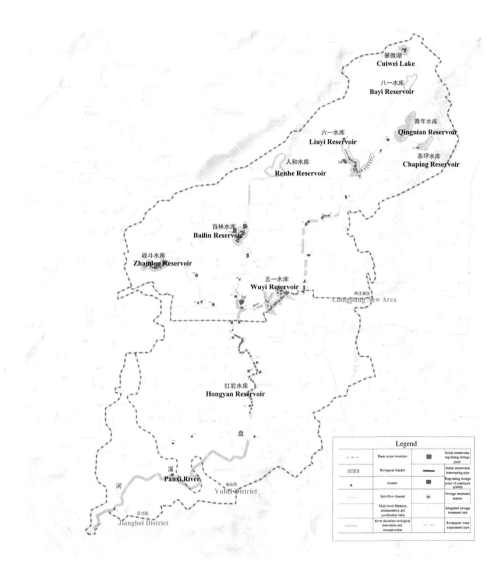

Figure 6.24 General layout for integrated treatment of the Panxi River Basin.

with black and odorous water (slightly/moderately black and odorous). There were also sewage direct discharge outlets and rainy-season overflow outlets along the river banks, and the river channel was of poor ecological landscape effect, arousing strong public resentment. The integrated water environment treatment was desperately needed to purify the water, eliminate black and odorous water and restore the healthy aquatic ecosystem.

In 2013, the integrated water environment treatment of 56 lakes and rivers was included into 25 major practical tasks for people's livelihood of Chongqing, and eight lakes in the Panxi River basin except for the Cuiwei Lake and Renhe Reservoir were included. Liangjiang New Area and Yubei District paid high attention to them and carried out specific measures such as intercepting pollutants for about 7.9 km pipe network in the basin, cleaning up about 15,900 m^3 silt in the basin, cultivating about 40,000 m^2 plants and cleaning more than 2,800 t garbage. The water environment in the basin was improved to a certain extent, but the Panxi River still lacked integrated treatment and supervision from the level of basin, especially the control of combined systems overflow pollution and initial stormwater pollution, resulting in difficulty in eliminating black and odorous water phenomenon and failure in ecosystem restoration and reconstruction.

In 2016, after the implementation of the Panxi River basin treatment scheme, benefiting from various environmental works such as source control, pollutant interception and initial stormwater control, the basin water environment was obviously improved with abundant aquatic plants at the river bottom. The treatment archived preliminary effects and the black and odorous water was basically eliminated. The service value of the basin water environment ecosystem was high after the treatment, and the treatment engineering archived good ecological benefits and obvious eco-environmental impact benefits. The water environment quality was improved and the water environment ecosystem was more secure and stable.

According to the evaluation requirements, places such as enterprises, communities and business districts within the radius of 1 km from the Panxi River were selected as the public review scope. The investigation team randomly distributed questionnaires to 133 citizens within the review scope of the Panxi River, among which 117 questionnaires were valid and 16 were invalid. The invalid ones mainly referred to those without real personal information filled or with many signs of alteration. According to the valid questionnaire results, 76 of them were very satisfied with the treatment results, 36 satisfied and 5 dissatisfied. Among the 116 valid questionnaires, 112 were satisfied with the Panxi River treatment results, with a satisfaction rate of 96%, so it was considered that the water body treatment achieved preliminary effects.

During the treatment process, some urgent problems were found out: the urban secondary and tertiary pipe networks were not complete and faced with leakage and blocking problems due to out of repair for long years; early treatment works stressed the reduction of pollution entering rivers and primarily focused on source control and pollutant interception, but the efforts for ecological restoration were inadequate, and measures for building a healthy, complete and stable river water ecosystem should be taken; the Panxi River basin already had relevant systems, such as a sound river chief system, treatment effect evaluation mechanism and pollution discharge licensing system to define the institutions, expenses, systems and responsible persons for daily maintenance and management of water bodies and various pollution treatment measures, but the implementation of various systems still needed to be intensified for so many departments involved without a collaboration mechanism.

For subsequent water environment treatment in the Panxi River basin, the following suggestions are put forward: (1) the unified design, unified systematic pollution treatment and unified scheduling for implementation are recommended for water environment treatment in the Panxi River basin; (2) it is recommended to adopt the implementation pattern of staged implementation and "treating exogenous pollution first, then endogenous pollution"; (3) it is recommended to tighten the pollution discharge supervision (domestic sewage direct discharge, industrial wastewater) in the basin; (4) it is recommended to combine the flood control works with water conservancy scheduling; (5) the urban river treatment shall pay attention to the continuous observation, monitoring and improvement after the treatment, for example, regular water quality monitoring, maintenance of healthy growth of various aquatic plants in water and plants on river banks and establishment of a long-term mechanism for water protection by residents. (6) The combination of engineering and ecological water purification and landscape construction in urban basin water environment treatment is sustainable and effective. For example, the construction of ecological river banks according to river morphology and functional demands, setting of ecological floating bed on river surface, planting of vegetation on river banks and construction of riverside parks are all feasible and effective landscape ecology measures.

Chapter 7

Epidemic prevention and control measures for urban drainage systems in mountainous regions

The urban drainage system is an important municipal public infrastructure, including drainage pipe network, urban sewage treatment plant, etc., which undertakes the important tasks of domestic sewage collection and treatment and plays an important role in reducing pollutants, ensuring the ecosystem safety and protecting urban water environment. During the COVID-19 outbreak, the novel coronavirus was detected in patients' feces, indicating the potential transmission and exposure way of the virus from toilets to septic tanks, sewage pipe network and sewage stations (plants). In order to ensure the safe and normal operation of urban drainage facilities, prevent the novel coronavirus from spreading via drainage systems and try to cut off links that may lead to human infection, the operating maintenance of drainage facilities shall be strengthened to ensure the cutoff of novel coronavirus transmission via drainage systems, so as to ensure the health safety of production staff for urban drainage systems, ensure the normal operation of urban drainage facilities and give full play to the role of "pollution control and anti-virus protection".

7.1 General protective measures for field operating personnel

7.1.1 General provisions

During the COVID-19 prevention and control, relevant operation management departments shall adopt all necessary protective measures to ensure the health security of production personnel during their on-duty period and carry out the safety protection training and education for production personnel to enhance their protection awareness and protective capability. Relevant operation management departments shall establish corresponding measures for request, use and management of protective equipment according to practical working environment and requirements and focus on the risks and protective equipment demands in different positions.

Relevant departments shall ensure sufficient anti-epidemic supplies in place and prepare enough masks, thermometers, hand sanitizers, disinfectant and other anti-epidemic supplies (the reserves of masks and disinfectant shall support the use for at least one week and be stored separately). As for workers who may have direct contact with sewage, two masks shall be prepared, one for use and one for backup, so that the mask wet by water can be replaced immediately.

The protective equipment used by field operating personnel during working shall meet relevant product quality standards and shall be collected in a unified way after use and disposed as domestic waste after thorough disinfection. The selection and specific use of disinfectant should be implemented by reference to *Guidelines for Disinfectant Use*. Besides, chemical disinfectant shall not be used to disinfect the air (space) when someone is present; glutaraldehyde should not be used for environment disinfection by wiping or spraying; high-concentration chlorinated disinfectant (available chlorine concentration >1,000 mg/L) should not be used for preventive disinfection; and ethanol should not be used for large-area spraying disinfection.

7.1.2 Protection of production personnel[15]

7.1.2.1 Operation and maintenance personnel of drainage pipes and channels

The operating personnel shall wear working or protective clothing, masks, goggles, latex gloves and other protective equipment in the whole process of above-ground inspection and work and try to avoid contact with sewage in drainage pipes and channels, avoid touching faces with their hands during working and wash their hands immediately after inspection or completing tasks for a certain period. If data recording is needed, relevant personnel shall regularly disinfect pens and other relevant stationery or bring their own stationery.

The underground work is not recommended during the epidemic outbreak. If it is unavoidable, the operating personnel shall wear protective equipment such as protective coverall, isolated-type respiratory protective equipment and latex gloves. After the work, the operating personnel and protective equipment used shall be strictly disinfected. The tools or equipment that have come into contact with sewage or sludge, such as washing truck, sewage suction truck, mud bucket, latex gloves, protective coverall and isolated-type respiratory protective equipment shall be flushed first and then subject to disinfection on site by spraying chlorinated disinfectant with an available chlorine concentration of 500 mg/L. The operating personnel shall use alcohol-based hand sanitizer or alcohol pad for hand disinfection. Fixed places for personnel and equipment disinfection shall be established if conditions permit.

7.1.2.2 Operation and maintenance personnel of sewage plants and stations

The zone management shall be implemented in sewage plants, and clean zones shall be set up in locations far away from sewage treatment process for employees to rest and eat. Before entering the clean zones, personnel shall be thoroughly disinfected and put clothes that have been exposed to pollution in pollution zones.

The operating personnel shall try to avoid direct contact with sewage, sludge, screenings and effluent of screenings during working. If it is unavoidable, they shall wear working or protective clothing, latex gloves, masks, goggles and face shields, and not touch their faces with their hands in the whole process, and shall wash their hands immediately after working or completing tasks for a certain period.

If the equipment disassembly is involved for equipment maintenance and overhauling, the equipment shall be subject to spraying disinfection first. As for the manual maintenance and disassembly of parts that have come into contact with sewage and sludge, the operating personnel shall wear working or protective clothing, two layers of gloves (nitrile gloves inside, rubber or cotton gloves outside), masks, goggles and face shields in the whole process.

As for the technology and equipment inspection in plants and stations, the working personnel shall wear working or protective clothing, masks, disposable gloves, goggles and face shields in the whole process. If data recording is needed, relevant personnel shall regularly disinfect pens and other relevant stationery or bring their own stationery. If conditions permit, the plants or stations shall provide front-line operating personnel with shower rooms, so that the personnel can wash thoroughly after working and before leaving the plants or stations, thereby blocking the diffusion path.

7.1.2.3 Laboratory testing personnel

Laboratories shall be kept clean and be ventilated for at least three times a day, 20–30 minutes each. The protective management on fume hoods and refrigerators shall be strengthened. The ventilation frequency can be reduced on foggy and hazy days.

Laboratories shall be regularly disinfected. Tables, office supplies and documents shall be disinfected with 75% ethanol solution every morning and evening, while telephone sets and mobile phones are recommended to be wiped with 75% ethanol solution by employees themselves, twice a day, or more times a day if they are used more frequently. The floors and toilets shall be disinfected by spraying chlorinated disinfectant solution with an available chlorine concentration of 500 mg/L. Laboratories shall strengthen the laboratory sample management, and during sampling or testing, relevant laboratory personnel shall wear working or protective clothing, latex gloves, masks, goggles and face shields, and not touch their faces with their hands in the whole process, and shall wash their hands immediately after working or completing tasks for a certain period.

If data recording is needed during the sampling or testing process, relevant laboratory personnel shall regularly disinfect pens and other relevant stationery or bring their own stationery. The laboratory wastewater and waste laboratory samples shall be thoroughly disinfected and then disposed according to original methods.

7.2 Management and control measures for urban drainage systems

7.2.1 General provisions

On the premise of ensuring normal operation of urban drainage systems, the outdoor manual operation shall be reduced as much as possible, and mechanical operations and automatic sampling shall be increased to ensure the health and safety of production personnel and also ensure unblocked urban drainage systems around designated medical institutions, temporary quarantine places and communities once with infected cases to avoid sewage overflow.

Proper disinfection measures shall be taken for urban drainage facilities to cut off the transmission of novel coronaviruses via urban drainage systems. The monitoring and prevention for designated medical institutions and temporary quarantine places shall be enhanced to eradicate the discharge of untreated wastewater or treated waste-water not meeting standards directly into urban drainage networks.

7.2.2 Drainage sources

7.2.2.1 Communities and public buildings

The key prevention and control areas shall include wash basins, toilet stools, floor drains, bathtubs, mop tubs, condensate drain pipes of air conditioner, water seal units at the connection between kitchens and grease traps, balcony down-pipes, etc. in communities and public buildings. The *Guidelines for Office Buildings to Deal with "Novel Coronavirus" Operational Management Emergency Measures* (T/ASC 08–2020) shall be referenced for the key inspection contents.

As for communities with infected cases, before the communities are released from quarantine, the stormwater and sewage from the communities' internal drainage systems shall not be discharged into municipal pipe network until being disinfected before the node connecting to the municipal pipe network e.g., septic tank outlet, community stormwater system's access point to the municipal pipe network).

7.2.2.2 Septic tanks

The key prevention and control areas shall include public septic tanks around designated medical institutions and temporary quarantine places and septic tanks in communities once with infected cases. As for the inspection and operations in buildings and communities with existing infected cases, the manhole covers and septic tank covers shall be inspected to make sure they are closed; second, various non-emergency clearing operations for community and public septic tanks shall be suspended. If the manual operation is necessary, the personnel shall be properly protected. Besides, warning lines shall be set around septic tanks with a protection distance no less than 10 m, and the protection distance shall be appropriately increased if buildings are downwind of septic tanks. The spraying disinfection shall be carried out twice a day for surfaces of inspection wells/holes and water outlets of septic tanks and areas within 2–5 m around them or flexible shields soaked with high-concentration disinfectant shall be covered on septic tank outlets.

7.2.2.3 Gutter inlets

The gutter inlet management and law enforcement shall be enhanced to eradicate the discharge of sewage and waste debris into gutter inlets. The debris and sludge cleaned out from gutter inlets shall be sealed during transportation and properly disposed.

7.2.2.4 Administrative licensing

As for the application for drainage license, the online application and offline mailing should be adopted during the novel coronavirus outbreak to ensure the online service quality and efficiency and try to reduce the manual sampling and monitoring for drainage license supervisory testing. If the manual operation is necessary, necessary personnel protective measures shall be taken.

As for the application for ditch connection and change, the online application and offline mailing should be adopted to ensure the online service quality and efficiency. While carrying out the ditch connection and change works, the construction personnel shall take necessary protective measures to avoid direct contact with sewage.

7.2.3 Drainage pipes and channels

7.2.3.1 Key prevention and control areas

The key prevention and control areas shall include drainage facilities (including drainage pipes and channels, drainage inspection wells and other ancillary facilities), sites with the risk of sewage overflow and combined overflow outlets around designated medical institutions, temporary quarantine places and communities once with infected cases.

7.2.3.2 Inspection and operation

During the epidemic outbreak, the pipe general inspection such as inspection by opening covers or entering pipes and channels shall be suspended. On the premise of ensuring the personnel safety and unblocked pipes and channels, the conventional pipe and channel dredging shall be primarily carried out by mechanical and hydraulic means, and manual dredging shall be avoided as much as possible. If the manual operation is necessary, necessary personnel protective measures shall be taken. The solid waste cleared out must be transported by sealed transport vehicles to places meeting relevant provisions for final disposal.

Relevant departments shall pay special attention to the inspection on the discharge of medical wastewater from designated medical institutions into drainage pipe networks and closely monitor the water quality of sewage discharged from hospitals into municipal pipe networks to ensure that medical wastewater is not discharged into municipal pipe networks until it meets the *Discharge Standard of Water Pollutants for Medical Organization* (GB18466-2005) and to prohibit the discharge of untreated or substandard treated medical wastewater into municipal drainage pipe networks; special attention shall be paid to the inspection on unclosed sections of drainage pipes and channels transiting from steep slopes and slow slopes in drainage zones where communities were once with infected cases. If temporary closing measures can be adopted, they shall be timely closed, while if temporary closing measures are hard to be adopted, warning signs shall be set up to prevent the public contacting with aerosols that may be formed by the water impact; special attention shall also be paid to the inspection on discharge of sewage from temporary quarantine places into drainage pipe networks, and specially assigned persons shall be arranged to be responsible for the disinfection of temporary quarantine places with integrated emergency sewage treatment facilities to ensure that sewage meets relevant treatment standards as specified in the *Technical Scheme for Emergency Treatment of Novel Coronavirus Polluted Medical Sewage* and then transported to sewage treatment plants for re-treatment and to prohibit the direct discharge or leakage of sewage to external environment; efforts

shall be made to check if there are sites with the risk of sewage overflow around designated medical institutions, temporary quarantine places and communities once with
infected cases. If any, warning signs shall be set up immediately to avoid the public
from direct contact; inspectors shall be arranged for daily inspection on sites with
the risk of sewage overflow and in case of any problem discovered, timely report to
the superior and handle according to emergency repair procedures; the sludge generated from sewage treatment facilities shall be treated as hazardous waste by reference to the *Technical Scheme for Emergency Treatment of Novel Coronavirus Polluted
Medical Sewage (Trial)*, which shall be subject to the unified disposal by institutions
with qualification for hazardous waste treatment and disposal; as for temporary quarantine places without integrated emergency sewage treatment facilities, the sewage
shall be transported via pipe networks to sewage treatment plants for treatment after
being disinfected by effective measures; the sewage overflow shall be eradicated and
chlorination devices shall be added upstream of overflow sites for temporary addition of chlorinated disinfectant with an available chlorine concentration no less than
10 mg/L. Enough disinfection time shall be ensured and the free residual chlorine shall
be no less than 0.5 mg/L.

Relevant departments shall draw sewage discharge routes from hospitals, temporary quarantine places and communities once with infected cases to sewage treatment
plants within their jurisdiction and arrange inspectors for daily inspection on drainage well lids along the routes to ensure no missing, loose or displaced well lids and unblocked sewage discharge routes. The management and control on combined overflow
outlets shall be tightened. Warning signs shall be set up to avoid the public from direct
contact, and chlorination devices shall be added upstream of overflow sites for temporary addition of chlorinated disinfectant with an available chlorine concentration
no less than 10 mg/L. Enough disinfection contact time shall be ensured and the free
residual chlorine shall be no less than 0.5 mg/L.

While handling emergencies such as sewage overflow or pipe collapse, emergency
repair personnel shall take necessary protective measures. The emergency repair shall
be carried out primarily by mechanical and hydraulic means to reduce the personnel
contact. In principle, the underground manual operation shall be avoided unless in
emergency circumstances. If it is unavoidable, it shall be approved by main leaders
of corresponding institutions and strict protective measures shall be taken first. The
criterion for relevant handling results is to meet basic drainage demands.

7.2.4 Pump stations

7.2.4.1 Key prevention and control areas

The key prevention and control areas shall include deodorization systems of pump
stations and all production areas in pump stations that may come into contact with
sewage, screenings and effluent of screenings.

7.2.4.2 Production inspection

As for the deodorization systems of pump stations, the deodorization system perfec-
tion shall be inspected to ensure the odor is emitted at a high altitude after being

suctioned and multi-level treated. As for all production areas in pump stations that may come into contact with sewage, screenings and effluent of screenings, the direct contact with domestic sewage, screenings and effluent of screenings in pump stations shall be avoided as much as possible during inspection. After all production areas in pump stations that may come into contact with sewage, screenings and effluent of screenings are cleaned up, the chlorinated disinfectant with an available chlorine concentration of 500 mg/L shall be used by reference to the *Guidelines for Disinfectant Use* for spraying disinfection of walls, floors, equipment, devices, garbage cans (areas) and surrounding environment, at least respectively once in the morning and afternoon, so as to ensure the environmental sanitation in production areas. If there are changing rooms in pump stations, the chlorinated disinfectant with an available chlorine concentration of 500 mg/L shall be used by reference to the *Guidelines for Disinfectant Use* to disinfect changing rooms, once before 8:00 AM every day. Screenings and effluent of screenings shall be sealed during transportation and appropriate disinfection measures shall be adopted.

7.2.5 Sewage treatment plants[16]

7.2.5.1 Key prevention and control areas

The key prevention and control units shall include incoming sewage pretreatment unit (screen room, aerated grit chamber), biochemical unit, disinfection unit, outgoing water and sludge treatment and disposal unit, overflow outlet prior to plant and deodorization system.

7.2.5.2 Production inspection

The incoming water quality shall be primarily inspected by automatic sampling and online testing, and close attention shall be paid to the residual chlorine concentration change in incoming water, and if the plant conditions permit, residual chlorine online monitoring equipment can be added. On the premise of ensuring the premise of effective protection, the monitoring on incoming water sources of sewage treatment plants shall be enhanced by increasing the water quality automatic testing frequency, and the technological parameters shall be timely adjusted according to changes in water quality and volume, so as to ensure the normal operation of sewage treatment facilities. On the premise of ensuring the safety, the drainage sump chamber prior to plant shall be operated at a low water level as much as possible. Besides, the intermediate lift pump station should be operated at a low water level.

As for pretreatment units, protective covers shall be added in front of equipment (screen, grit-water separator, etc.) easily causing sewage splashing. Besides, the site inspection on pretreatment units shall be reduced to try to avoid direct contact with screenings and grit, on which disinfectant shall be timely sprayed.

As for biochemical units, the testing frequency of sludge concentration in biochemical basin shall be reduced, and the activated sludge concentration shall be closely tracked primarily by video monitoring inspection, online testing and necessary site visual inspection. In case of any abnormality, the instrument data shall be timely compared and testing shall be conducted; if reduced sludge activity is found, the residual activated sludge discharge should be reduced to increase the sludge concentration and extend the actual operating sludge age. Meanwhile, the aeration rate shall be appropriately increased to give priority to the stable operation of aerobic biological treatment unit.

As for disinfection units, the disinfection shall be enhanced practically, and several disinfection measures, such as by adding disinfectant, ozone, or using ultraviolet radiator, shall be adopted according to actual operations of sewage treatment plants to ensure the disinfection processes are effective and the fecal coliform count indicators meet the requirements as specified in *Discharge Standard of Pollutants for Municipal Wastewater Treatment Plant*, so as to effectively control the sanitary risk of outgoing water. In areas seriously affected by epidemic outbreak, sewage treatment plants implementing Class I(B) standard shall consider increasing the disinfectant amount if conditions permit to try to meet the values of Class I(A) standard.

The outgoing water quality shall be inspected primarily by automatic sampling and online testing and the number of manual sampling times shall be decreased as appropriate. If the manual sampling is necessary, the operating personnel shall take strict protective measures. The outgoing water quality testing shall be enhanced and the frequency of fecal coliform count testing shall be increased from weekly to daily, while other routine tests may be conducted daily as determined by corresponding departments according to actual conditions. Close attention shall be paid to outgoing water quality data and the monitoring on fecal coliform count in outgoing water shall be intensified. If the fecal coliform count does not meet standards, the outgoing water disinfection system should be assisted with sodium hypochlorite or liquid chlorine disinfection to ensure the outgoing water quality meets standards, but after the dosing, attention shall be paid to the disinfectant's impact on aquatic life in receiving water. Besides, the monitoring on TN and TP indicators in outgoing water shall also be intensified and if the indicators do not meet standards, measures such as external carbon source adding or enhanced chemical phosphorus removal should be taken to ensure the biological treatment system's capacity for nitrogen and phosphorus removal.

As for sludge treatment and disposal units, the frequency of sludge organic content monitoring shall be reduced. The sludge moisture content shall be monitored weekly but shall also be strictly controlled to ensure smooth sludge discharge, and the sludge dewatering room shall be ventilated and disinfected. The dewatered sludge shall be cleared on the same day of being produced and shall be properly treated, and shall be further treated by such means as heat drying or lime alkaline stabilization as much as possible if the sewage treatment plant conditions permit. The sludge not timely transported shall be treated by reference to *Technical Guideline for Hospital Sewage Treatment*, according to which the sludge storage pond shall be used as the sludge disinfection pond for enhanced sludge disinfection with lime or bleaching powder as the disinfectant, and the time of sludge storage in plants shall not be more than 5 days. The supervision on sludge disposal shall be enhanced to implement the quintuplicate form system for sludge transfer and standardize the whole process of sludge collection, transportation and disposal. The sludge shall be sealed during the storage and transportation, and if necessary, disinfectant shall be sprayed properly and relevant records shall be made to avoid the sludge pollution transfer. The system of "specially assigned vehicles, persons, routes and sites" shall be implemented for transporting sludge to unified disposal sites, the whole-process supervision shall be conducted on the safe disposal of sludge. Moreover, urban domestic sludge disposal enterprises shall be strictly prohibited to receive and dispose of sludge from medical wastewater treatment system without authorization.

Overflow outlets prior to plants shall be blocked off to prohibit sewage overflow. If it is difficult to block the overflow outlets, existing combined system overflow pollution control facilities shall be fully utilized. If combined system overflow rapid purification facilities are constructed, the regulation and control shall be enhanced to improve the removal, disinfection and disposal of sediments and suspended solids. If there is no purification facility, chlorination devices shall be added upstream of overflow outlets for temporary addition of chlorinated disinfectant with an available chlorine concentration no less than 10 mg/L. Enough disinfection contact time shall be ensured and the free residual chlorine shall be no less than 0.5 mg/L.

As for deodorization systems, the air suction volume shall be improved to increase the ventilation rate of odor generation unit and ensure the collected odor is discharged at a high altitude after being treated and meeting discharge standards.

7.2.5.3 Rural sewage treatment facilities

As for rural sewage treatment facilities with disinfection units, relevant requirements of sewage treatment plants shall be implemented for the facilities' operation and maintenance. As for rural sewage treatment facilities without disinfection units, the disinfectant added shall be increased as appropriate in villages and towns already with epidemic situation to improve the disinfection effects, but at the same time, attention shall be paid to the disinfectant's impact on aquatic life in receiving water.

7.2.5.4 Other sewage treatment facilities

As for small sewage treatment facilities such as integrated sewage treatment facilities, relevant requirements of sewage treatment plants shall be implemented. Low construction standards are implemented for a few small sewage treatment facilities and for facilities without advanced treatment facilities such as disinfection facilities, the disinfectant dosing facilities can be added as appropriate to improve the disinfection effects, but at the same time, attention shall be paid to the disinfectant's impact on aquatic life in receiving water.

7.2.6 Reclaimed water reuse facilities

During the novel coronavirus outbreak, if reclaimed water is used as landscape river water, warning signs shall be set up, and multiple measures such as chlorination, UV and ozone disinfection shall be taken to provide multiple disinfection technological guarantees to ensure the normal operation of recycling and disinfection technology, so that *The Reuse of Urban Recycling Water—Water Quality Standard for Scenic Environment Use* (GBT 18921-2002) can be complied with and the sanitary risk of reclaimed water utilization can be controlled. Besides, other reclaimed water utilization methods that may have potential human exposure shall be suspended and strict preventive measures shall be developed to avoid drinking by mistake, misuse and misconnection.

7.3 Guarantee measures

7.3.1 Strengthen organizational guarantee

Relevant municipal departments for health and ecological environment shall establish a mechanism for joint prevention and control to promote information sharing and ensure information disclosure, so that operating agencies of urban drainage facilities can timely and accurately acquire relevant information about the distribution of designated medical institutions, temporary quarantine places and communities once with infected cases and medical wastewater treatment, so as to co-establish response plans for drainage system emergencies and make joint efforts to win the battle of epidemic prevention and control.

Operating agencies of urban drainage facilities shall establish epidemic prevention and control safe production working groups responsible for establishing regulations on safe production during epidemic prevention and control period to guide, supervise and publicize the epidemic prevention and control work of urban drainage systems. The safe production responsibility system shall be strictly implemented, and emergency disposal mechanisms and daily production reporting systems for epidemic prevention and control period shall be established and improved. Personnel shall be assigned on duty by turns for 24 hours and leaders shall be designated for each shift. In case of abnormality during epidemic outbreak, emergency measures shall be timely adopted and relevant information shall be reported to local competent authorities.

7.3.2 Ensure material supply

Relevant departments shall list operating agencies of urban drainage facilities in the business directory with necessary priorities and include them into the important material transportation guarantee management system during emergency periods. The supply of production materials such as chemicals, equipment and apparatus, and protective materials such as masks, goggles, gloves and protective clothing shall be guaranteed by opening "green channels" for equipment maintenance, material supply and field maintenance to maintain smooth transportation of production and operating supplies, chemicals and sludge, so as to ensure normal operation of drainage systems and safety of personnel. Meanwhile, the guarantee for power and other factors of operating agencies of urban drainage facilities shall be enhanced to ensure normal power supply.

7.3.3 Intensify anti-epidemic management

Operating agencies of urban drainage facilities shall carry out necessary physical examinations on employees in accordance with relevant epidemic prevention and control requirements, establish the daily examination and reporting systems for the staff's health conditions, take the staff's temperatures when they are coming to and getting off work, properly record their health conditions and timely report any suspected case to local health departments for proper treatment. Besides, relevant provisions shall be strictly implemented and the closed-off management shall be implemented by sewage treatment plants to prohibit unauthorized access of external personnel to sewage treatment plants.

7.3.4 Enhance publicity and education

Relevant companies shall enhance the publicity and education to raise drainage facilities production personnel's awareness of biological hazards of sewage and shall also intensify training on correct use of protective equipment to enhance production personnel's awareness of safety protection.

Relevant departments shall enhance the publicity and education to call on residents not to throw masks, sanitary pads, plastic bags and other articles easily blocking sewers to guarantee unblocked sewage pipe networks and call on residents to highly value the family and personal hygiene and protection by reducing touching, frequent hands washing, promoting natural ventilation of homes and public washrooms and checking water seal protection of floor drains to avoid virus transmission via sewers.

Chapter 8

Results and prospects

8.1 Results

Water is the foundation of survival, source of civilization and essential of ecology. Water resource is a major controlling factor of ecological environment, while water eco-civilization is an important part and basic guarantee of eco-civilization. As of 2020, the ecological safety of important rivers and reservoirs is guaranteed and the goals of unobstructed rivers, clear water, green slopes and beautiful banks have been preliminarily achieved. In order to accelerate the construction of a place with beautiful mountains and rivers, Chongqing has insisted on "promoting well-coordinated environmental conservation and avoiding excessive development", given overall consideration to protection and development, systematically planned for "water, gas, noise and residues", comprehensively promoted the "construction, treatment, management and rectification" and resolutely carried out the pollution control work, and obvious results have been achieved currently. The water pollution control involves many industries, heavy workload and wide scope, and also involves work in many aspects, such as industrial structure optimization, cleaner production level promotion, long-term stable operation of treatment facilities and monitoring supervision. Over the last few years, Chongqing has achieved overall improvement in the water pollution control work through concerted efforts of government regulation, enterprise follow-up, social organizations and the public engagement.

In March 2018, Chongqing, based on comprehensive inspection results, announced that the black and odorous water had been already basically eliminated in 48 black and odorous water segments in Chongqing, with the public satisfaction all over 90%, and the water quality of 56 lakes and reservoirs had also been improved in 2017. Meanwhile, Chongqing also required all districts and counties to strengthen measures such as water circulating and supplement, ecological restoration and "sponge city" construction. From 2018, the requirements for "sponge city" construction indicators shall be fully implemented in all construction links of urban water treatment projects to strictly control the stormwater runoff pollution, so as to improve the urban water environment from the source.

In order to further improve the water environment quality, Chongqing has established the *Work Program for Integrated Water Environment Treatment in Key Basins of 2019* and specified five measures, focusing on promoting the solving of prominent environment problems of substandard key rivers to ensure continuous improvement of water environment quality. Chongqing plans to establish a recording and cancellation management system for key treatment basins to urge the cause analysis, source analysis and measure establishment of districts and counties where Laixi River, Linjiang River, Binan River and rivers with water quality inferior to Class V make a list of treatment projects and quantify treatment projects and measures to realize defined goals, tasks, deadlines and responsibilities. Chongqing Ecology and Environment Bureau will implement the supervision according to specific schedules, regularly schedule the work progress and water quality and will take measures such as inquiry, supervision according to specific schedules, regional restricted approval or supervision by municipal leaders, for those with slow progress or poor treatment effects.

The service value of the basin water environment ecosystem has been improved after the treatment, and the treatment engineering has archived good ecological benefits. As for mountainous Chongqing in the Three Gorges Reservoir Region, the water environment quality has been improved and the water environment ecosystem has become more secure and stable. In the Yangtze River basin, except for a few river segments, the rate of water quality reaching relevant standards in water functional areas, Grade-I and Grade-II areas, has been significantly improved, indicating that preliminary effects have been achieved through improving water pollution control and water resources protection in recent years.

8.1.1 Speeding up water ecological civilization construction

In recent years, Chongqing has taken the water ecological civilization city construction as an important platform base and practical action for promoting construction of beautiful Chongqing. Focusing on improving water environment quality in the Three Gorges Reservoir Region, Chongqing has systematically controlled pollution sources to comprehensively prevent and control pollution and comprehensively strengthen the eco-environmental protection in the Yangtze River Economic Belt. Chongqing has highly valued sponge city construction and taken the advantage of the first-batch national sponge city construction pilot city for the vigorous planning and construction of a sponge city with mountainous characteristics, orderly guiding and promoting the construction of mountainous sponge city. Chongqing has successively set up the Chongqing Municipal Engineering Technology Research Center of Sponge City Construction and Chongqing Sponge City Construction Expert Committee, and under the guidance of Chongqing Municipal People's Government and competent authorities has carried out the planning, design, standard establishment and engineering construction for Chongqing sponge city construction, forming a series of documents for sponge city construction policies and technical indicator systems and establishing the point-to-area sponge city construction pilot pattern of "1 (national-level Yuelai New Town) + 3 (municipal-level Wanzhou District, Bishan District and Xiushan County)" covering the whole municipality. Chongqing has actively promoted the implementation of the strictest water resources management system by giving play to its water environment advantages, insisting scientific planning, promoting from the leadership, innovating mechanisms and making concerted efforts.

8.1.2 Significantly improving ecological environments of rivers and lakes

Water is a controlling factor of ecosystem, and the benign cycle of water is a basic guarantee of healthy ecology and an important foundation of good life. The ecological basin construction is to realize the benign water cycle through a series of measures, so as to make water adapt to human society. From aspects of water quantity, water quality and stormwater utilization, its scientific connotation can be summarized as "flood sponge, black and odorous water cleaning and stormwater resource recycling", namely, "to reduce sewage discharge of social water cycle into natural water cycle by efficient utilization; give priority to the layout of green infrastructure (such as mountain, water, forest, farmland, lake and grass) and blue infrastructure (river, lake, wetland, etc.) and the utilization of soil reservoirs and groundwater reservoirs; reasonably arrange surface gray infrastructure (namely man-made reservoir, dike, pump station, sewage treatment plant) and integrate with new advances in modern information technology to realize the multi-process (surface, soil and underground) joint regulation and control of water quantity, water quality, silt and water ecology, realize the de-extremalization (minimize drought and flood disasters) to the greatest extent and systematically solve water problems.

Since the release of the No. 1 general river chief order of Chongqing, under the effects of actual implementation of four-level river chiefs, effective execution of municipal departments, proper implementation of district and county Party and governmental departments and orderly participation of the public and, through the "three-stage" work of comprehensive investigation, centralized treatment and effect solidification and improvement, a total of 4,055 illegal, direct and random sewage discharge problems have been identified in Chongqing, 3,345 of which have been rectified, with a rectification rate of 82.5%. The illegal and random sewage discharge in Chongqing has been effectively curbed, and records have been made for direct discharge problems to accelerate the solving. The water environment quality has been improved and the river and reservoir messy phenomena have been obviously improved, indicating that remarkable effects have been achieved through implementation of the general river chief order. For instance, in 2019, 93.4% of the 211 cross sections in Chongqing met the water functional requirements, up 7.6% from a year earlier. Among them, the proportion of cross sections with the water quality reaching or superior to Class III in the 42 cross sections included in the national assessment was 97.6%, up 7.1% from a year earlier.

8.2 Prospects

From review on the development of Chongqing environmental protection cause over the past 70 years, especially since the reform and opening-up after the founding of People's Republic of China, the ecological civilization construction has been promoted steadily and the environmental protection has developed orderly, showing continuous achievements in pollution control and constantly improving the ecological environment. During the period, Chongqing, on the basis of protecting water environment in the Three Gorges Reservoir Region, has comprehensively promoted the pollution prevention and control work, accelerated the construction of a place with beautiful mountains and rivers and practically built an important ecological barrier in the upper reaches of the Yangtze River, showing continuous effects in the eco-environmental protection.

In the past few years, Chongqing has actively issued a series of integrated treatment measures regarding water environment and made significant breakthroughs in the pollution prevention and control, making significant contributions to improving ecological environmental quality and enhancing ecological civilization construction and environmental protection.

However, the water environment treatment is a long-term task. In order to make further achievements in prevention and control of pollution and protection of clear water, Chongqing will keep moving forward on the road of water environment treatment and look forward to the future, full of hope.

8.2.1 Guiding ideology

In the future integrated water environment treatment in Chongqing, new development concepts shall be firmly established and effectively implemented; the guiding spirit of "giving priority to eco-environmental restoration of the Yangtze River" and "promoting well-coordinated environmental conservation and avoiding excessive development" shall be implemented; the ideas of "giving priority to water saving, spatial balance, systematic treatment and joint efforts of government functions and market mechanism" shall be adhered to; the "five must-nots" baseline shall be strictly observed; the "protecting water resources, controlling shorelines, improving water environment, restoring water ecology and realizing water safety" shall be taken as primary missions; the river chief system shall be comprehensively implemented in all rivers and reservoirs of Chongqing; the river/reservoir management protection mechanism characterized by clear responsibilities, orderly coordination, strict supervision and effective protection shall be established to provide an institutional guarantee for building an important ecological barrier in the upper reaches of the Yangtze River, maintaining healthy life in rivers and reservoirs of Chongqing and realizing sustainable use of river/reservoir functions, so as to make Chongqing a place with beautiful mountains and rivers.

8.2.2 Basic principles

In order to prevent and control pollution, four principles must be adhered to in integrated water environment treatment of Chongqing:

1. Insist on "ecology in priority" and green development. The rehabilitation of rivers and reservoirs shall be promoted and the ecological functions shall be maintained by firmly establishing the concept of respecting, conforming to and protecting nature, properly handling the relationship between management and protection of rivers and reservoirs and their development and utilization, building a strong green development background, strengthening planning constraint and insisting on requisition-compensation balance.
2. Insist on leadership of the Party and government and departmental cooperation. A working pattern of level-to-level administration and implementation at all

levels shall be formed by establishing and improving a core responsibility system centered on the Party and government leadership responsibility system, defining responsibilities of river chiefs at each level, improving departmental collaboration mechanism, intensifying work measures and coordinating forces of all parties.

3. Insist on problem-orientated and addressing both symptoms and root causes. Prominent problems for management and protection of rivers and reservoirs shall be properly solved by making overall plans for the upstream and downstream, both banks, main streams and tributaries, inside and outside reservoirs, implementing the "one strategy for one river" and "one strategy for one reservoir" and treating water and land pollution simultaneously according to actual conditions of different regions and different rivers and reservoirs.

4. Insist on tightened supervision and strict evaluation. A good atmosphere where rivers and reservoirs are cared about and protected by the whole society shall be created by controlling and managing water in accordance with laws, establishing and improving systems for management, protection, supervision, evaluation and accountability of rivers and reservoirs and expanding channels for the public engagement.

8.2.3 Goals and conclusions

We shall ensure the harmonious coexistence between human and nature and must establish and practice the concept of "lucid waters and lush mountains are invaluable assets" and adhere to the basic concepts of resource conservation and environmental protection. Establishing and practicing the concept of "lucid waters and lush mountains are invaluable assets" have become strategic thinking for promoting ecological civilization construction and coordinating the relationship between environmental protection and economic development in the new age. The "lucid waters and lush mountains" refers to the combination of natural resources and ecological environment, while the "invaluable assets" refers to the aggregation of human material wealth. For establishing and practicing the development concept of "lucid waters and lush mountains are invaluable assets", ideologically, we need to deeply understand that natural ecology is valuable, and the protection of nature is a process of adding value to natural value and natural capital. In order to make Chongqing a place with beautiful mountains and rivers and win the tough battle of pollution prevention and control, Chongqing has put forward the overall goals of "one not-inferior-to, three only increases" for integrated water environment treatment in the future.

1. To ensure the water quality of the Yangtze River main stream is not inferior to that of the inflow water, and the lengths of other rivers reaching water quality goals of the water functional areas only increase.

2. To ensure ecological basic flows in river channels only increase.

3. To ensure water areas of all rivers and reservoirs in Chongqing only increase.

In the future, all departments at all levels in Chongqing will accelerate strengthening weak links in environmental protection by improving sewage treatment, solid waste disposal, road traffic pollution control, vessel and wharf pollution control and eco-ecological protection supervision big data. On the aspect of improving environmental supervision level, Chongqing has defined the ecological protection red line of $20,400 \, km^2$, accounting for 24.82% of its total area. Chongqing will strictly implement the environmental assessment system, reinforce environmental supervision, carry out special activities, implement full coverage of centralized environmental supervision and urge the full performance of environmental protection responsibilities of all departments by means of inquiry, supervision according to specific schedules, etc. Meanwhile, Chongqing will pay attention to the application of scientific and technological means to daily management of ecological environment, build and put into service of Chongqing ecological environment big data platform to preliminarily form an ecological environment "network" that connects the Ministry of Ecology and Environment and district (county) bureaus of ecology and environment vertically, links governmental departments at all levels horizontally and links enterprises and other nodes externally.

Besides, Chongqing will also regularly evaluate the water quality improvement effects in key basins and timely increase monitoring points for rivers with cross sections of substandard or deteriorated water quality and their major tributaries and carry out water quality monitoring on main streams and tributaries locally or comprehensively.

Meanwhile, Chongqing will set up a team to conduct normal supervision and guidance on key basins. Full-time personnel will be transferred to carry out supervision and guidance on pollution control of districts and counties where the Laixi River, Linjiang River, Binan River and other major key rivers with water quality inferior to Class V are located in and will focus on supervise and check if the district or county basin treatment list is completed on schedule, if expected results are achieved and if pollution sources have prominent illegal, random and direct discharge problems, and put forward reasonable treatment advice to the district or county government according to the supervision situation.

Moreover, Chongqing starts to establish a water quality data consultation system for key rivers and will conduct irregular comparative analyses on data obtained from water quality automatic stations and manual monitoring and hold consultations primarily for sharp concentration changes of main pollutants at cross sections, river water quality class degradation and abnormal water quality of water from Sichuan, Guizhou and Hubei to research and put forward solutions and rectification measures.

Chongqing will provide technical support for water environment treatment of key basins through multi-party cooperation. Taking the opportunity of the ongoing resident follow-up study on ecological environmental protection and restoration of the Yangtze River, Chongqing will include the treatment of substandard rivers into the resident follow-up study and invite Chinese Research Academy of Environmental Sciences, relevant municipal experts together with district or county governments to research the scientificity, systematicness, pertinence and reasonability of key basin treatment measures, so as to guide districts and counties to further deepen and optimize integrated treatment schemes for rivers.

Chongqing will speed up the ecological civilization system reform, carry out environment credit rating, implement the basin horizontal ecological protection compensation mechanism reform and ecological environmental damage compensation system reform and deepen the reforms of "streamline administration, power delegation and service optimization".

In order to fully promote the integrated water environment treatment progress and build a beautiful Chongqing, it is believed that in the future integrated water environment treatment, Chongqing will gradually form a water environment treatment system dominated by the government, primarily relying on enterprises and jointly participated in by social organizations and the public, and Chongqing will achieve the grand goal of building a place with beautiful mountains and rivers step by step.

8.2.4 Construction of "four banks of two rivers"

Cities always complete the self-renewal step by step in the continuous development and progress. Taking the riverside public space in Yangpu District, Shanghai, as an example, as a cradle of China's modern industry, the riverside space in Yangpu has witnessed over 100-year industrial development history in Shanghai. In west part of the riverside public space, there is an old water plant with a history of nearly 140 years— Yangtszepoo Water Works, which is the earliest modern water plant in China and is still in use now.

Nowadays, you can find scenery everywhere in this public space. The riverside "stormwater garden" was a "stormwater depression" vulnerable to waterlogging, now it is built into a "small but beautiful" urban landscape by stormwater recycle system according to local conditions. The 5.5 km-long Yangpu waterfront area, which was a production shoreline mainly with factories and warehouses, has been transformed into an ecological landscape shoreline primarily decorated by park green spaces suitable for daily leisure activities. The past industrial rust belt has been changed into a beautiful leisure space. Obviously, Yangpu waterfront is a typical case of city renewal.

The city construction, like a living being, is a process of dynamic renewal. As for Chongqing, which also has a long river shoreline, the development experience of Shanghai is enough to serve as a model for the development of Chongqing. Chongqing has very strong regional characteristics, and the collective symbol formed by the culture accumulated from the history of Chongqing and the characteristics of unique social and cultural environment can constitute the urban space capital of Chongqing itself. Active use of the unique urban space capital can create a better future for Chongqing.

Chongqing is a famous city of mountains and rivers, and the "four banks of two rivers" in its main urban area is the main axis of urban development, a core area of the "land of natural beauty" in Chongqing and also an important gateway and carrier for Chongqing to promote inland opening-up. The "four banks of two rivers" refers to the river shoreline of the Yangtze River and Jialing River. The "four banks" line is 394 km-long in total and the hinterland along the rivers is nearly 231 km^2. Riverside roads built in the last century in Chongqing main urban area mainly focused on traffic functions. However, with people's increasing demands for leisure and landscape, more public activity and waterfront leisure spaces, as well as cultural landscape, have become key focuses for the riverside construction. Therefore, the "four banks of two rivers" needs new planning, and the landscape, traffic, cultural and economic aspects will be gradually considered for the development of riverside roads. Moreover, the shifting of focus from vehicle traffic to pedestrian traffic will provide citizens and tourists with more ecological waterfront public spaces to let them have a better understanding of the urban construction and historical and cultural deposits of Chongqing.

The construction and improvement of the "four banks of two rivers" are the epitome of Chongqing's urban renewal. According to Chongqing Municipal People's Government's unified deployment of the future urban improvement action plan, Chongqing will accelerate the treatment and improvement of the "four banks of two rivers" and strive to build the "four banks of two rivers" into an urban model of "a land of natural beauty, a city with cultural appeal" with the goal of building *ecological belts with beautiful mountains and rivers, vertical urban landscape belts, convenient shared recreation belts and cultural customs belts,* making the beautiful scenery with "mountains, rivers, cityscapes and bridges" become a city name card. It is expected that by 2022, the "four banks of two rivers" will basically complete the treatment and improvement, reaching the quality level of "international first-class riverside belts" and becoming a city name card displaying "a land of natural beauty, a city with cultural appeal".

8.2.5 Construction of "lucid water and green banks"

The water pollution control involves the public vital interests and well-being. Chongqing has always adhered to the concept of "lucid water and green banks" and will continue to make coordinated efforts to promote whole-basin treatment of secondary rivers in its main urban area, so as to reproduce the phenomenon of lucid water, green banks and fish gliding in the limpid deep water for citizens to enhance their sense of gain, which is also the ultimate goal of urban water environment improvement, to further improve the urban living environment quality.

During the construction of "lucid water and green banks", Chongqing will stick to principles of "one strategy for one river" and "water pollution control, water ecological restoration and water resources protection" ("water treatment from three aspects"), build a sponge city, implement the whole-process stormwater and sewage discharge control and comprehensively promote the "water treatment from three aspects" while respecting the basin characteristics and adapting measures to local land and water conditions. The "lucid water and green banks" treatment and improvement works will be implemented in steps to gradually improve the river ecosystem. Sticking to the people-oriented, people's livelihood-focused and people-centered principles, Chongqing will intensify the leisure, recreation and sightseeing functions of shorelines and enhance the accessibility, participation and comfort of spaces to make it more convenient for citizens to play in waterfront areas.

The river treatment will continue to be subject to overall planning and orderly implemented. Whole-basin treatment and improvement schemes and work plans will be made and orderly implemented in steps by following the concepts of integrated "sewage treatment plants, drainage networks, rivers, lakes and banks" and whole-basin treatment and based on the overall consideration of the upstream and downstream, main streams and tributaries, both banks, ground and underground and water bodies along shorelines. Besides, a long-term mechanism will be established to continuously improve the hydrodynamic conditions and water purification capacity by keeping ecological improvement and long-term effect maintenance.

In order to achieve the overall goals of "lucid water and green banks", the requirements for future river pollution treatment are as follows: the river water quality meets requirements for Class IV or above water as specified in Environmental Quality Standards for Surface Water; main river segments never dry up throughout the year and maintain stable ecological basic flows; urban built-up areas with the basins with an area of $394 \, km^2$ basically meet requirements for sponge city planned indicators and the green coverage rate of green buffer zones reaches more than 80%. A long-term water environment treatment mechanism will be established and improved to realize the long-term clear water of urban water bodies; the leisure, recreation, fitness and other functions of shoreline public open spaces will be intensified and the accessibility, participation and comfort of these spaces will be enhanced to make it more convenient for citizens to play in waterfront areas.

In order to promote smooth implementation of "lucid water and green banks" treatment and improvement actions, a mechanism for "government supervision, market operation, performance evaluation and pay for performance" will also need to be established and improved to promote a market-oriented mechanism–construction–operation and maintenance mechanism with specialized social capital as the main body for "lucid water and green bank" construction. The whole-basin operation and maintenance enterprises shall timely disclose the operation and maintenance information of each river to accept the public supervision and reporting.

It is expected that by 2021, Chongqing will build 20 rivers with "lucid waters and green banks" with a total length of 427 km. In order to achieve the goal of "lucid water and green banks", Chongqing not only needs to protect and treat water resources and water ecology but also needs to improve the waterfront landscape and supporting facilities to build water bodies into clear rivers that we can see fish gliding in the limpid deep water, open shared green corridors and comfortable and pleasant ecological spaces.

8.3 Guarantee measures

The integrated Three Gorges Reservoir Region basin treatment has achieved phased results, but the environment carrying pressure is still large and the water environment stability still faces a huge challenge. The water environment pollution control is a long-lasting battle and it is recommended to start from the following aspects:

8.3.1 Transforming management functions of government sectors

8.3.1.1 Defining government's responsibilities in urban water environment treatment

The government functions shall be transformed as soon as possible by taking advantage of the reform of state institutions to clarify the boundary between governments and market and society, define functions that must be performed by governments and functions that can be transferred to the market and society, so that the public services can be transformed from the mode of "being direct offered and managed by governments" to the mode of "being purchased, evaluated and supervised by governments", then the government sectors can focus on establishing development planning, strengthening supervision and management, knowing about the public demands, etc.

In the urban water environment treatment process, main functions of governments are as follows:

1. To establish policies and planning. Focusing on scientific and reasonable planning of water control infrastructure, governments shall establish a unified and appropriately advanced water treatment planning design and management standard for

the whole city; strengthen the formulation of laws and regulations on water supply, drainage and sewage treatment; research and establish relevant provisions for water supply and drainage and water pollution control; specify restraint mechanisms of "water supply after drainage" and "water supply determined based on drainage" and establish institutional systems compatible with integrated water supply, drainage and sewage treatment.

2. To improve integrated supervision and law enforcement mechanism and perfect water environment monitoring network. Governments shall make full use of information technology to accelerate the construction of ecological environmental monitoring network; construct a three-dimensional water environment monitoring system based on the "point monitoring" of key sources, risk sources and sensitive sources, "linear monitoring" of basin water systems and "plane monitoring" of satellite remote sensing and improve environmental law enforcement systems, integrate and set up integrated law enforcement teams for eco-environmental protection, so as to implement law enforcement for eco-environmental protection in a unified way.

3. To improve the system of linking administrative law enforcement with environmental judicature. Governments shall establish systems for information sharing, case notification and business training for each other among public security departments, procuratorates, courts and environmental departments to form a strong joint force of environmental law enforcement and intensify sanctions and penalties for ecological and environmental illegal and criminal activities.

8.3.1.2 Defining distribution of subject power and responsibility

The division of labor and distribution of power and responsibility of all subjects shall be defined to specify their respective resources mastered and scope of responsibility. In case of difficulties in supply, they shall cooperate with each other to face together; in case of problems to be solved, they shall not pass the buck to each other to shirk responsibility.

The transactional work for urban water environment treatment shall be centralized by stripping water-related transactional functions of governmental sectors and assigning them to one responsibility subject. On the basis of defining functions of relevant functional departments, governmental functions of relevant departments such as departments for environment protection, water, development and reform, economy and information, finance, land resources, housing and urban-rural development, administrative law enforcement, traffic, agriculture, forestry and marine fishery shall be integrated by stripping water-related transactional work undertaken by them. Municipal state-owned enterprises are supported to integrate water-related resource assets for centralized operation and management. A closed chain for water infrastructure investment, construction and operation integrated with water supply, drainage and sewage treatment shall be established to implement integrated infrastructure for municipal and county (district) water supply, drainage and sewage treatment, promote continuous improvement of water ecological environment quality in Chongqing and constantly expand and optimize urban development space. Constraint mechanisms for economic responsibility auditing, information disclosure and recourse shall be improved.

As for other urban water environment treatment projects, such as river flotsam cleaning, profitable enterprises can be attracted to participate in the treatment by such means as franchising, commission contract and leasing or volunteer teams and environmental protection organizations can be invited to participate in the treatment.

8.3.2 Promoting development of social organizations

8.3.2.1 Improving voluntary service system

The supervision cost can be effectively reduced through public supervision. The river chief system and lake chief system shall be promoted and implemented by official account publicity and supervision as well as carrying out volunteer service activities. Volunteers shall be selected from the public to serve as nongovernmental volunteer supervisors for water environment treatment, who shall cooperate governments at all levels to implement the "river chief system", regularly patrol and inspect river channels, timely clean waste in rivers and prohibit river-related construction, sand excavation, dumping rubbish into rivers or building farms near rivers.

8.3.2.2 Improving the professional level of environmental social organizations' personnel

Various social organizations and people committed to environmental protection shall be guided to participate in environmental protection work legally and orderly to give full play to their positive roles in guiding the public concern and support for environmental protection. The support for environmental social organizations shall be increased by providing more policy support and assisting in solving their difficulties in office space, fund raising, etc. Environmental social organizations shall be encouraged to cooperate with water pollution treatment environmental protection enterprises and water environment scientific research institutions to organize environmental protection professional skills training for relevant personnel of environmental social organizations, so as to improve their professional skills in water pollution monitoring and water environment investigation and improve the team quality and capability.

8.3.3 Improving fund management mechanism

1. A comprehensive budget management system shall be established, with the fund revenue and expenditure management as the main clue and construction goals as the core. Based on the comprehensive budget management system, the management level shall be improved by such management means and methods as improving two fund management processes, intensifying three internal control and restriction mechanisms, implementing control of three risk points.

    a. Improving organizational structure.
    A budget management committee shall be established and "measures for comprehensive budget management" shall be developed. The budget management committee shall, on the principles of "whole-process control, efficiency priority, expenditure according to revenue and defined power and responsibility",

organize relevant departments and personnel to predict budget goals, research, coordinate and review various budgetary matters.

b. Enhancing fund budget management

Annual goals and tasks shall be broken down to detail budgetary revenue and expenditure items. Fund budgets shall be made and the budget enforcement shall be analyzed, and the performance evaluation shall be intensified.

c. Intensifying three internal control and restriction mechanisms

In order to further ensure effective implementation of fund management work, the actual management shall focus on three internal control and restriction mechanisms, namely, the organizational control mechanism: each business link shall be assigned to corresponding department and responsible person and incompatible functions shall be separated; authorization approval mechanism: the approver's way of authorization, authority, procedure, responsibility and relevant control measures shall be defined; restriction and supervision mechanism: an internal restriction system and post responsibility system shall be established to specify the responsibility and authority to ensure mutual restriction and supervision in work.

d. Improving approval processes for fund management

While the fund budget management is enhanced, approval processes for fund management shall be continuously improved, mainly including the approval process for financial revenue and expenditure and approval process for project payment.

e. Implementing risk control

According to possible risk points in construction in progress, price settlement and payment and contract management, corresponding risk control measures shall be made, with the core to minimize the impact of unexpected future events on fund management, mainly including the construction in progress risk control, contract risk control and project price payment risk control.

2. A third-party fund supervision system shall be established to improve the fund risk prevention capacity. The sewage treatment facility construction fund management is the core of project management in a sense, because all other management items, such as safety, quality, project progress and civilized construction can't do without fund support. Based on the importance of fund management, a project company shall establish a third-party fund supervision system to further improve the fund risk prevention capacity and ensure special construction fund for special use. Therefore, a fund supervision system shall be established to specify the responsibility and improve the supervision process.

3. The financial supervision shall be carried out to intensify dynamic fund management and make use of professional advantages of the financial supervision to establish the concept of changing passive control to active control. Through the whole-process and all-round investment control working method of financial supervision before, during and after the event, the goal of dynamic fund management can be archived.

Bibliography

1. Xiaoling Lei, Bo Lu, *Sponge City of Mountainous Region: Theory and Practice* [M], Beijing: China Architecture & Building Press, 2017.
2. Bo Lu, Xiaoling Lei, *Sponge City of Mountainous Region: Case Study* [M], Beijing: China Architecture & Building Press, 2017.
3. Xiaoling Lei, Bo Lu, *Black-odorous Water Treatment & Management in the Three Gorges Reservoir Region: Theory and Practice* [M], Beijing: China Architecture & Building Press, 2018.
4. JCH [2014] No. 275, *Technical Guideline for Sponge City Construction: Low Impact Develop Development Stormwater System Construction (Trial)* [S].
5. JZH [2016] No. 18, *National Building Standard Design System for Sponge City Construction* [S].
6. CJJ83-2016, *Code for Vertical Planning on Urban and Rural Development Land* [S].
7. GB50513-2009, *Code for Plan of Urban Water System* [S], 2016.
8. GB50180-93, *Code for Urban Residential Areas Planning & Design* [S], 2016.
9. CJJ37-2012, *Code for Design of Urban Road Engineering* [S], 2016.
10. GB50420-2007, *Code for the Design of Urban Green Space* [S], 2016.
11. CJ/T 340-2016, *Planting Soil for Greening* [S].
12. GB50014-2006, *Code for Design of Outdoor Wastewater Engineering* [S], 2016.
13. CJJ6-2009, *Technical Specification for Safety of Urban Sewer Maintenance* [S].
14. DBJ50/T-293-2018, *Technical Specification for Green Space of Sponge City* [S].
15. *Guidelines of Chongqing on Urban Drainage System Risk Prevention during Novel Coronavirus Pneumonia Outbreak (Trial)*, Chongqing Municipal Research Institute of Design & Chongqing University, 2020-02.
16. *Guidelines on Rural Drainage System Risk Prevention and Control during Novel Coronavirus Pneumonia Outbreak (Trial)*, Rural Environment Scientific and Technological Industry Alliance, 2020-03.

Index